More from the reviewers ...

The end of the 20th century reveals the growing uncertainties with regard to the future course of the world economy, and doubts regarding the very purpose of the homo economicus in his technology-driven environment. We need more humanistic approach to our heavily incorporated life and this book provides us with an attractive option.... The book will be widely read by all those concerned about future developments in human resource management, corporate management, Asian business studies, microeconomics, business ethics, comparative business studies and related disciplines. Highly recommended.

— Prof. Chris Czerkawski, Chair, International Business Finance,
Hiroshima Shudo University, Japan

This book is vintage Raymond Kao! He has directed his exciting, creative and innovative mind towards solving some of the critical problems and stumbling blocks of the future as we revitalize and re-create how we do business in the world! This is a revolutionary work that will stimulate deep thought on how we must shape the businesses of the future to maximize employee participation, leadership and achievement. Don't pass up this masterpiece of future thought if you are a corporate executive struggling with how to re-create your firm for survival and achievement in the future, or if you are a neophyte future business leader dreaming of how you will create the next wonder company of world business, or if you are a teacher of business who is willing to stimulate your students to think beyond the mundane thoughts and ideas of the past on how to create, organize, and run companies.... This book should be "absolutely must" reading for all who claim to be *in/en*trepreneurs in large firms. The philosophy can change mere managers to entrepreneurial leaders. We all talk about values, social responsibility, ethics, employee motivation, future-driven management, etc. This book provides the vehicle to pull this all together in a methodology that can work and truly make the firm a means to benefit overall society, and itself. These are thoughts and concept that will help to solve the serious problems we have in business today.

— Prof. W. Fred Kiesner, Professor of Management and Entrepreneurship,
Loyola Marymount University, US

About the author

Raymond **W. Y. Kao** is currently Shaw Foundation Professor at Nanyang Business School, Nanyang Technological University, Singapore; International Consultant and Modern Management Training Specialist to the United Nations Industrial Development Organization's Programme to Promote Women's Participation in the Modernization of China; Editor-in-Chief of the *Journal of Small Business and Entrepreneurship*, Canada; Strategic Advisor to the Chinese Chamber of Commerce Institute of Management Singapore; and Special Advisor to the School of Business and the Centre of Entrepreneurship and Economic Development, Centennial College, Canada. His other involvements include serving as a member of the Editorial Board of *internationales gewerbeauchiv*, Switzerland; a member of the Small Business Consultative Committee to the Minister of Small Business, Government of Canada; President of the International Council for Small Business (based in the US); Professor of Entrepreneurship, University of Toronto, Canada; and first Tasman Fellow, University of Canterbury, New Zealand.

His main publications include *Entrepreneurship: A Wealth-Creation and Value-Adding Process*, *Small Business Management* (third edition); *Entrepreneurship and Enterprise Development*; and *Accounting Standards Overload: Big GAAP Versus Little GAAP*.

A Canadian born in China, and educated in China, Canada and the UK, he has been listed in the Canadian *Who's Who* since 1986.

An Entrepreneurial Approach to Corporate Management

Raymond W. Y. Kao

Shaw Foundation Professor
Nanyang Business School
Nanyang Technological University

PRENTICE HALL

Singapore   New York   London   Toronto   Sydney   Mexico City

First published in 1997 by
Prentice Hall
Simon & Schuster (Asia) Pte Ltd
317 Alexandra Road
#04-01 IKEA Building
Singapore 159965

© 1997 Simon & Schuster (Asia) Pte Ltd
A division of Simon & Schuster International Group

Prentice Hall, Simon & Schuster offices in Asia: *Bangkok, Beijing, Hong Kong, Jakarta, Kuala Lumpur, Manila, New Delhi, Seoul, Singapore, Taipei, Tokyo*

Library of Congress Cataloging-in-Publication Data

Kao, Raymond W. Y.
  An entrepreneurial approach to corporate management / Raymond W.Y. Kao.
       p.   cm.
  Includes bibliographical references and index.
  ISBN 0-13-626772-6 (pbk.)
  1. Entrepreneurship. 2. Corporate culture. 3. Social responsibility of business.   I. Title.
HB615.K3628 1997
658.4'22-dc21   97-1495
                                                                   CIP

Printed in Singapore

5 4 3 2 1   01 00 99 98 97

ISBN 0-13-626772-6

Simon & Schuster (Asia) Pte Ltd, *Singapore*
Prentice Hall, Inc., *Upper Saddle River, New Jersey*
Prentice Hall Europe, *London*
Prentice Hall Canada Inc., *Toronto*
Prentice Hall of Australia Pty Limited, *Sydney*
Prentice Hall Hispanoamericana, S.A., *Mexico*

Contents

AN ENTREPRENEURIAL APPROACH TO CORPORATE MANAGEMENT
Structure of the Book

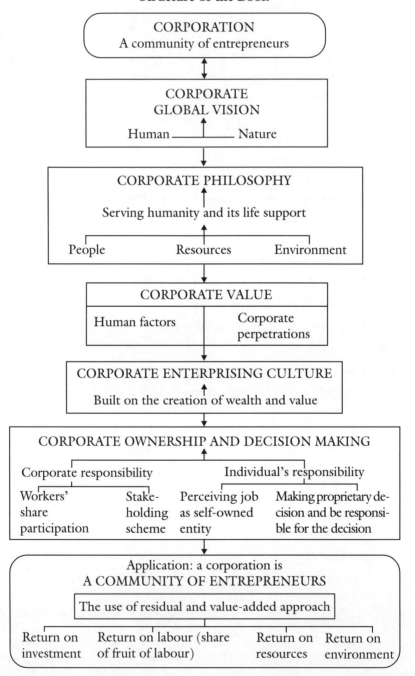

Preface

Recently, I was sitting in the Singapore Zoo with my wife and son, watching the lions pacing outside, while I relaxed in air-conditioned comfort, sheltered from the heat of the blazing tropical sun. Suddenly, the thought came to my mind that like the lions outside, I too was in a prison. I shared this thought with my son, who replied, "Dad, it has been said before that we are prisoners of our own technology." I realized the truth of this, but also knew that, unlike the lions, we have the choice to step outside our prison, if we so prefer. We remained inside by our own desire, because it was too comfortable.

This very same thought made me think about the rate of return on investment (ROI), three letters to which the engines of our economy, the stock exchange and the theory of the firm are wedded. ROI is also our prison, and a powerful one; it has captured business management education and practice in its entirety, in a lifetime sentence with virtually no chance of parole. Of course, why would we want to leave that air-conditioned shelter?

Unlike technology, which constantly advances at an almost breathtaking pace, the walls of the business management prison are virtually static; we still live in the "midsummer night's dream" of the 1950s and early 1960s, and there is not a hint of expanding the limits of knowledge. Of course, some claim that we have had a few jail breaks, but if we scrutinize it carefully, there were not many breakthroughs, but out of the kind heart of the prison warden on duty, prisoners (for example, corporate managers whose performance measurement is based on their ability to satisfy ROI requirement) are merely allowed to step out of their cells to take a breath of "recycled old prison air".

Consequently to break out from ROI prison is not going to be an easy task. In the first place, it has perhaps the best structured prison one can imagine, with a team of professional trained wardens (such as, the accountants, economists, stock exchange commission) watching over from the top that would make jail break next to impossible. But if the prisoners are willing and determined, there are possibilities. On the other hand, like the air-conditioned shelter story, if the prisoners are comfortable in jail, why should they want to break out? A jail break only happens if those making a living in the ROI prison cell realize that corporate life is more than just making profit for others, and freedom is of greater value than confinement. Nevertheless, in order to make a breakout effective, tools are needed – tools such as "environmental accounting", and "models for the integration of resources, environment and the value of people into the theory of firm". True, "environmental economics" has been the subject of research and model-building for quite some time, but the business community does not react to economic analysis, unless the accountants also take action to improve the way they measure costs and disclosure practice about environmental damage, among other things, the destruction of human assets, and loss to the society. This is unlikely; "no risk, no gain, no risk, no loss" is the golden rule for accounting, and breaking it is to risk the profession's reputation, is it not?

Like the accounting practice, business management is also averse to change; they are very proud of their "profit making" tradition and "ROI" performance. Why not? Relying on "invisible hands" has made living in unimaginable luxuries possible. Moreover, it seems everything is going well. There are more millionaires in the world than ever before, why bother to change?

But the poor are getting poorer and their numbers are increasing all over the world; meanwhile the rich are getting richer, enjoying luxuries beyond comprehension. Both the US and Japan experienced serious economic setbacks in the middle of the 1990s. The setbacks were due to rapid capital accumulation that requires persistent high rate of return on investment, and failure to accommodate capitalists' wants that force corporations to cut back and downsize their operations (with beautiful words such as re-tooling and re-engineering). These, compounded with investment outflow for reasons of seeking greener pastures (countries where labour and resources are cheap and plenty, and lesser or no environmental legislations) elsewhere led to an economic slowdown, and subsequent recession causing job loss. To unemployed individuals, the continuing joblessness would demoralize their spirits, cause them to lose their pride, confidence, and sometimes dignity. The increasing number of beggars in the streets of large cities is a case in point.

How could things be better? For a start, we should not just focus on money, and let the "invisible hands" worry about rewards and punishment. Like it or not, the old system is failing, and it will be seen first where failure can be afforded the least – in the developing countries. Sir John Houghton, a renowned expert in climatology, predicted that global warming will have massive effects on water and food supplies in these areas (*The Straits Times*, Singapore, 17 February 1996, p. 9). Forests will die, malaria and other diseases will run rampant and starving refugees will wander from country to country as the weather turns more and more extreme. But this is not the fault of the native peoples of the developing countries. Global warming is caused by man-made methane emission from the burning of fossil fuels. For the culprit we must look first to the industrial nations, where car manufacturers push for high production volume and more and more sales. Co-operation between people and between nations is vital to solve this dilemma, and indeed there has been progress in this regard – an example is the European Union. But what else do we see in Europe? The French President Jacques Chirac refused to give up "nuclear superiority" and conducted nuclear weapons testing in the South Pacific despite near universal international condemnation. The Spanish telecommunication giant Telefonica and French Canal Plus pay-television company circumvented European legislation on monopolies. Of course, there are countless other incidents and occurrences, and they will continue to happen because we have not yet faced up to the reality that our "invisible hands" have no feelings and no compassion. The environment will certainly not be able to defend itself, nor can our unborn children. They cannot complain or form unions, or go to war, even though they have every right to, as our silent partners.

We want our children, we want humanity to continue to have a place on earth. We nurture our children, provide them with love and care, set up college funds, and try to maximize their financial inheritance; yet we have very little or no concern for the life support they need. In the name of profit, we mercilessly take what is supposed to belong to our children and children's children for our present consumption. What a tragedy it is to see advertisements saying "We bring the future to you for your present enjoyment." If we are to be put on trial by our unborn descendants, would we stand a chance? Of course, if we do not change our ways, a better question may be: Who would be there to judge us, if there is going to be the "end of birth"?

It is to avoid this judgement that I have written this book; not to defend the wrongs of the present, but to admit guilt and try to make some wrongs into rights. To do this, we must be freed from the ROI prison in which we have put ourselves (I hope without landing in a far less comfortable one). Capitalism, for better or worse, has been instrumental in making the world what it is today. This book begins by questioning the basis of capitalism, starting with profit, ideology, education in the business ("B") schools and how we govern corporations.

It is not the intention of this book to be merely critical, or simply provocative. To replace this system, the later chapters go on to suggest that corporations must be considered as social entities; responsible to all their contributors. The top-down style of management must be replaced by a bottom-up approach, taking into consideration individual cultural values and integrating them into a corporate enterprising culture. Corporate managers should go ahead and make all the "profit" they want, but not at the expenses of other people's misery, or through the waste of scarce resources or the deterioration of the environment. Environmental accounting is introduced and a value added approach to corporate performance measurement is illustrated. Resources scarcity dictates that unless there are new approaches and wasteful mismanagement ends, we will end up eating grass roots and leather boots, and be thankful for that.

While of course I do not have all the answers, it is my intention to plant a few seeds of thought in the readers' minds. I have had the privilege of enjoying the shade of trees that others before me have planted; it is my hope that the seeds which I sow, will germinate to provide shade for others.

An Entrepreneurial Approach to Corporate Management

In writing this book I am indebted to countless people. I would like to first express my admiration for Roger Wolf, former Dean of the Faculty of Management at the University of Toronto, Canada, who clearly told a financial supporter of the Faculty that he could keep his money, rather than violate the former's principle of academic integrity. I would also like to express my gratification for the people in the film industry who "put social issues before the box office" (to use the words of actor Christopher Reeves in his presentation at the 1996 Academy Awards ceremony); in today's world, the movies reach far more people than books ever could, and those who would use them to send important messages should be applauded.

The manuscript of this book was formally and informally reviewed worldwide by a number of people. A religious reviewer met me in Switzerland in September 1996. He sincerely said to me: "Raymond, I like what you wrote, I believe the good Lord likes it too. For what you have done, I am sure you will go to heaven." I didn't know how to respond for a moment; then I said to him: "When I get there, I'll send you my greetings through e-mail."

On a more personal note, I am grateful to my wife, Flora, for her tolerance of my 3.00 a.m. wake up times, her help and inspiration. I would like to thank my sons, Rowland and Kenneth, for their help with this manuscript, as well as Mr Tan Wee-Liang, the director of the Entrepreneurship Development Centre at the Nanyang Technological University, Singapore, and other friends for their encouragement. Finally, I would like to thank the reviewers for their time and efforts, in particular their comments and valuable suggestions that made this book possible.

Raymond W. Y. Kao

1

Introduction: A corporation is a community of entrepreneurs

A corporation is not a money-making dream machine for the interests of the few, but a community of entrepreneurs created for the purpose of generating wealth for the individual and adding value to society.

Corporation: The Money-making Dream Machine in the Market Economy

What is a corporation? To most of us, the answer is as simple as "a legal entity, incorporated under the law to do business". But there are variations. According to the *American Heritage Dictionary*, it is "a body of persons granted a charter legally recognizing them as a separate entity having its own rights, privileges, and liabilities distinct from those of its members." *Black's Law Dictionary* (6th edition, 1990) defines a corporation as: "An artificial person or legal entity created by, or under the authority of, the laws of a state ... The corporation is distinct from the individuals who comprise it." To make it simple, *The Devil's Dictionary* describes a corporation as: "An ingenious device of obtaining individual profit without individual responsibility." When the same question was posed to the author's ten-year-old daughter, she unhesitatingly answered in her original fashion: "If I have a corporation, I will let the company make money for me." When she was asked how she was going to make money, she said: "I would go to a lawyer to create the corporation, then I would hire an accountant to keep the books, and I would go to the company once in a while to collect money." In short, a corporation to her is a money-making dream machine in the market economy.

It may also come as a surprise that the whole world apparently looks upon the market economy as the beacon of economic growth. Some of us have even dedicated ourselves to advocate that the market economy should be "free" with no public intervention, but let "the invisible hand", the unbiased pricing system, govern supply and demand. The system, according to advocates, is a remarkable human achievement which forms the basis of democracy, entrepreneurship and anything that has to do with economic success. But if you look at it differently, if you remove all these beautiful words of glorification without public intervention, the real market economy that exists in the human environment is nothing more than gross animal behaviour. It has not got away (and never will) from the dominance of Darwin's doctrine: in the world of competition, the strong will live and the weak will die; thus, if you can't compete successfully in the marketplace and make yourself rich, you don't deserve to have "water, air or basic life support", death is inevitable. However, animals compete on their natural strength, they do not cheat (maybe they do, but at least not in the same way as in business, with well-prepared business plans and/or strategies), manipulate, corrupt or harm the same species as some corporations behave in the marketplace; yet we call our "almost corporate-dominated market economy" the attainment of human civilization, and even go as far as to link it directly with the noble ideologies of economic freedom and democracy.

The incredible part of a corporation is that it is a creature, with two pairs of godparents guiding its honoured path through the human environment, protected by its own indestructible infrastructure, unassailable even by acts of God. Throughout the world, a corporation as we know it is:

1. Protected by law.
2. Fed by accounting practice.
3. Nursed with milk and honey through the economists' theory of the firm.
4. Backed by the stock exchanges.

Note: (1) and (2) are the first pair of godparents, while (3) and (4) are the second pair of godparents.

A corporation works like a dream machine under the glorious umbrella of the "market economy" and "capitalism". This machine, over a period of time, has grown up to include the biggest monsters that mankind has ever made. Just look at the size of some corporations: Mitsubishi, IBM, Sony, General Motors, Standard Oil, ICI, BP and others. Henry Ford was right to say: "A great business is really too big to be human." However, a corporation is not human, but immortal, and it can expand its size indefinitely; the sky is the limit.

As we know it, both humans and corporations (including what they own) are governed by the law, but with a difference, because a corporation is not human. A person who violates the law is punishable by the law, when found guilty. Depending on the nature of the crime, in extreme cases, the person can be jailed for life or sent to the gallows. Once a person is in jail, there is very little interaction with the outside world. A corporate charter can be revoked or cancelled, but this is not comparable to putting a human to death. Even without a charter, a corporation can function through humans. Reinstatement is more often than not just a matter of time. In some circumstances it merely involves a name change which is certainly not the same as death for a person. How can a dead human be reinstated or revived from the dead just by a simple name change?

As a corporation is not human, it can virtually get away with anything that is considered to be "legal", and often gets around the legal system. For example, during the Japanese invasion of China in the early 1940s, after the Japanese Imperial Army conquered the resource-rich north-eastern provinces and placed the "last emperor" on the throne, Chinese people were used as laboratory subjects in experiments run in the interests of corporate gain. There was no law to prosecute the Japanese corporate executives who executed or used the Chinese civilians.

Many of us remember the Union Carbide accident which occurred on 3 December 1984. A cloud of the toxic chemical methylisocyanate escaped from a chemical plant in Bhopal in central India which spread across the poor quarters of the city. People living around the plant did not know what to do as the disaster unfolded. When they realized what was happening they ran towards the plant to seek help from the company, unaware that they were running to their deaths! Over 2,500 people died and it was estimated that some 100,000 others suffered serious injuries, including blindness and permanent lung damage. At that time it was considered to be the worst environmental disaster in industrial history. Following the accident, the company's management (the board of directors, CEO and other high level decision-makers), instead of attempting to compensate the community, the victims' families and the environment, proceeded to liquidate a substantial portion of the company's assets and give them out to shareholders in special dividends (Hawken, 1994, p. 116). What Union Carbide did was inhuman, but not illegal. There was no law that could place Union Carbide under arrest; not even the human behind the inhuman act could be punished. According to the *U.S. News and World Report*, 115 of the Fortune 500 companies were convicted of a serious crime during the 1980s, yet they continue to prosper. What about those who committed a crime but were not prosecuted or punished? Did these events happen in the name of profit for the corporations, or for the consumers? For economic prosperity or humanity?

Corporations are more intent on re-organizing the world to make it more comfortable for themselves, instead of for its inhabitants. Yet, everywhere we see signs marked with the ironic words "in the interest of our consumers".

Bottomline Driven Corporations

In the business world, "profit" and the "bottomline" both reflect the last line figure shown on the operating report or profit and loss statement. Therefore, more often than not, both terms are considered to be interchangeable. On the other hand, it is obvious that "profit" reported by itself can be meaningless, if it is not related to investment. It is much the same as the story about killing a housefly with a sledge hammer – it is effective, but is it efficient? The answer is obvious, it is not efficient, since it employs too much input (a sledge hammer) and yields too little output (killing a small housefly). A corporation whose profit and loss statement shows a profit of $10 million in a particular period, could mean little, if the investment used to earn that profit amounted to $10 billion. The rate of return on investment (ROI) for the period of undertaking is only 0.1%, in comparison with the normal interest rate of 5%. It is much the same as the hammer and fly story: too much input and very little output. Therefore, for corporations relying heavily on external equity financing, the bottomline or ROI approach to measuring the effectiveness and efficiency of corporate performance becomes evident.

In short, corporations, particularly those involved in raising funds from the public (through the stock exchange), and their managers, instead of being driven by profit, are driven by the bottomline or ROI. (See Figure 1.1.)

Figure 1.1   The management function in bottomline driven corporations

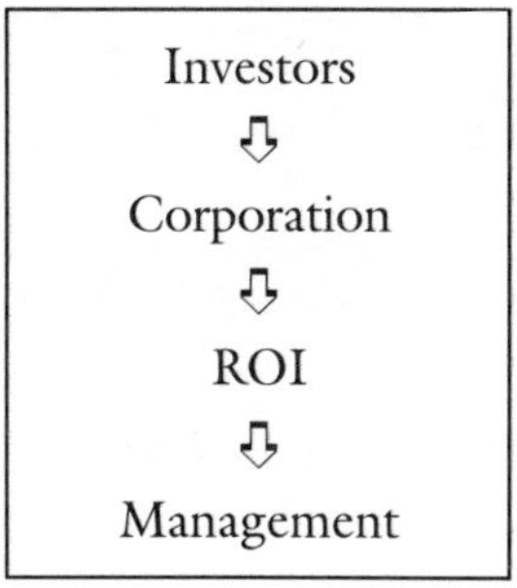

```
┌─────────────────────────┐
│       Investors          │
│           ⇩              │
│      Corporation         │
│           ⇩              │
│         ROI              │
│           ⇩              │
│      Management          │
└─────────────────────────┘
```

Note: Managers are striving for the attainment of desirable ROI through corporate activities to satisfy investors' expectations.

Profit Driven Corporation

A profit (or bottomline) driven corporation is an economic entity which pursues the making of profit through business transactions as it functions in society. As has always been said by business people: "profit is what we are in business for".

Unfortunately, with few exceptions, a large number of people in business, professionals and the academic community do not trouble themselves to make the distinction between short- and long-term profit, or question what profit really is, and how it is to be measured. Without exception, the challenge of profit determination is in the hands of accountants, and therefore, the profit driven corporation is in effect accounting profit driven. There is little evidence to indicate that the accounting profession has attempted to make distinctions between long-term and short-term profit, and only period profit (almost without exception one year) is used to summarize a corporation's performance. Therefore, a profit driven corporation is, in fact, a short-term (one year) accounting profit driven corporation. Consequently, corporate managers are pressured, and in effect motivated by their short-term accounting profit performance and measured by the same. (See Figure 1.2.)

Figure 1.2   The management function in profit driven corporations

A profit driven corporation could be an owner managed business, and/or family business where key decisions are made by a management group consisting of members of the family who are both financially and emotionally involved in the operation of the business. These corporations, if not reliant on external financing, may have motivations other than profit. Their objectives may include long-term growth, providing jobs for family members, providing for their children's future, etc.

The crucial challenge for corporate managers who are under pressure for short-term profit or ROI performance is the question of whether or not they can visualize the light at the end of the dark tunnel. For example, managers of either short-term profit driven or bottomline driven corporations may fail to appreciate the difference between long-term profitability and short-term financial profit. In this case, if their attention is focused on short-term accounting profit or ROI performance, even when the company has a brilliant success in new product development or an invention which is a true "technological breakthrough", if it is unable to attract investment to commercialize or make it a financial success, it could be a failure, in spite of, or even on account of the "invention" success. On the other hand, if someone steals the invention, commercializes it and becomes financially successful in the name of "entrepreneurship" – the thief would be an entrepreneur, and the inventor would be a failure. If this is the case, what then, is the purpose of management? Bear in mind that there are corporations whose decision-makers perceive their company to be not a piece of property, owned by its shareholders who have the freedom to do whatever they want, but a community, with obligations to its employees, customers and the surrounding environment, as well as to those who have provided it with finance. Profit is, of course, necessary and essential if the company is to survive and grow, but it is not an end in itself. Some German and Japanese corporations do have this type of business culture, and some US and Canadian companies may also have similar attitudes towards business. But the crucial challenge is to understand and determine how the individual manager is systematically motivated to behave in the marketplace. (See Table 1.1.)

Entrepreneurial Driven Corporations

An entrepreneurial driven corporation is an organization distinct from others as it is a community of entrepreneurs combining corporate (collective) and individual efforts to create wealth and add value to society. (See Figure 1.3.)

As shown in Table 1.2 all business entities can be entrepreneurial driven corporations, *as long as the corporate culture is guided*

Table 1.1 Perceived management cultural differences between "profit driven" and "bottomline driven" corporations

Area	Profit driven	Bottomline driven
Ownership	Venture founder/family members	Shareholders (outside investors)
Value	Venture founder's values, family values	Corporate value (difficult to define, normally appearing in the mission statement, containing catchwords such as: to serve the market needs, continuing work to provide opportunity for all in a free and competitive environment for profit)
Sources of finance	Owner/manager, family investment, love money, debt financing	Public financing, listed with the stock exchange, or venture capitalists
Managerial focus	Survival profitability, long-term profit	Bottomline, ROI, relatively short-term profit
Growth pattern	Vertical growth	Vertical growth and horizontal expansion
Managerial decision-making behaviour	Entrepreneurial inclined, action oriented	System oriented, staff control

An Entrepreneurial Approach to Corporate Management

Figure 1.3   The management function in an entrepreneurial driven corporation

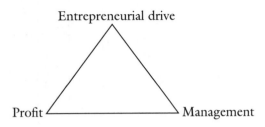

Entrepreneurial drive

Profit

Management

and directed to create wealth and add value in the interest of the individuals in the firm, the corporation (including investors) *and society.* Therefore, the challenge to be addressed is how to impact managers' entrepreneurial mindset to direct corporate policy and behaviour in the marketplace as communities of entrepreneurs.

The Power of the Corporation and Corporate Managers

Corporations possess virtually unlimited power to steer a nation and world economy but have no soul or mind of their own. Whether or not corporations function in society for the common good, to take from the poor and helpless and offer the rewards to the rich, or simply to be a mechanism to facilitate capital accumulation in the interest of a few is all in the hands of their managers. Corporate managers, as we know them, are key individuals in society exerting great influence on the well-being of mankind now and in the future. It is their decision who gets a job, who should be awarded the key to the executive washroom, and it is also their decisions and actions that satisfy shareholders' desire for ROI and affect the environment. Corporate managers for better or worse and directly or indirectly help to create the society that provides us with high standards of living. On the other hand, there are corporate managers and executives who, for whatever reasons, will do anything to achieve their goals, even at the cost of human lives, and the future of our children, as long as they consider it to be legal. After all, there are skilful legal counsellors ever willing to help.

Table 1.2 Perceived management cultural differences between "profit driven", "bottomline driven" and "entrepreneurial driven" corporations

Area	Profit driven (1)	Bottomline driven (2)	Entrepreneurial driven
Ownership	Venture founder/family members	Shareholders (outside investors)	(1)+(2)
Value	Owner/manager's personal values and/or family values	Corporation value (ROI or other wording contained in the corporate mission statement)	A community of entrepreneurs working for the common good and creating wealth for the individual
Source of finance	Essentially private, debt financing	Public financing, equity, debt (listed with stock exchange)	(1)+(2), (1) or (2)
Business growth pattern	Mostly vertical	Vertical growth and horizontal expansion	(1)+(2) or (1) or (2)
Managerial focus	Profitability	Bottomline, ROI	Creativity, innovativeness for growth and adding value to society
Managerial decision-making behaviour	Both long-term and short-term profitability	Satisfying shareholders' ROI requirement	(1)+(2) as the result, from wealth creation and value adding activities in the marketplace*

* Under the circumstances, profit is considered to be a meaningful measurement for businesses whose management appreciate the difference between accounting profit and relatively real profit under the "profit" concept perceived by economists. On the other hand, if a firm primarily seeks short-term profit, it would be considered the same as bottomline driven profit.

Who controls corporate managers?

In the old days, a business undertaking was managed and controlled by its owner. A corporation is a separate entity, usually controlled by its owners or shareholders, with managers hired to manage the operation. But this may not always be the case. What matters the most is that if shareholders of the corporation have no control over their managers, who then, controls the actions of the corporation's management? According to Henry Mintzberg (1984, p. 110): "In a democratic society, what justifies owner control of the corporation any more than worker control, or consumer control, or pluralistic control?" The corporation's board of directors, in effect, manages and controls the corporation. The theory of the firm, and the law of the state gives a corporation the status as an entity, while shareholders hold ownership, and every working individual in the corporation, including the CEO and all board members, are employees of the corporation. Hence, the ultimate management and control of a corporation should lie in the hands of the employees, even though the members of the board could be block shareholders (the managers) who are supposed to be working in the interest of all the shareholders. If they do not work in the interest of shareholders whose interest then, are they working for? Even more surprising is a study (rather old, but reflecting what happened at that time) which reported that of all the directors of Fortune 500 companies in 1977 only 1.6% represented significant shareholders' interests (Mintzberg, 1984, p. 110).

Either in theory or reality, it matters not how empirical results suggest who controls the corporation, but it is inevitable that the members of the board work in their own interests. A combination of the remuneration, dividends paid on their investment and stock value appreciation through corporate activities, express corporate performance in the form of ROI. It is the ROI that drives the corporation managers to function in the marketplace. The power of corporate managers is an expression of the human ability to conquer nature and use resources for the common good or in the interest of the few, or both. (See Table 1.3.)

Table 1.3 Corporate decision-making apparatus

Decision	Decision-making unit	Remark
Absolute authority (the authority can be manipulated by block voting and the use of proxy)	Shareholders' meeting held annually	If shares are widely distributed, the individual shareholder is of no great significance to the decision-making process
Highest (and effective) decision-making	Board of directors (powers are limited by law and corporate by-law. But the tenure for a director can in reality be a lifetime – only a few countries legally limit the directors' tenure in office, for example, in Singapore to three years)	Block shareholders have the winning edge under any circumstances
Strategic policy level decisions	Managers (including executives) below the board level may act independently	Key group, the dominant force in corporate management
Operational level decisions	Managers below the executive level	Having operational authority, decisions could have a direct impact on people, resources and the environment

The Tragedy for Bottomline Driven Corporate Managers is that They are Pressured to Earn Profit, but have No or Little Idea about Cost

Corporate managers, the decision-makers of corporate activities in the marketplace (and particularly those managers below the board level), are, regardless of their background or education, the movers and shakers of the world economy. There is very little doubt about their great contribution to "economic growth" and the prosperity of "capitalism". The tragedy is that whilst corporate managers will do what they can in the pursuit of profit and meeting the required ROI, they have little or no idea (or do not care) about the cost involved in earning the profit. Ideally, in order to determine profit, all costs must be accounted for, but this is not true in practice. All costs used in business transactions are derived from the accounting process, and use of the accounting method to measure cost is far from adequate. Because the accounting process is viewed as sacrosanct and accountants, as a whole, are at least publicly unwilling to risk their professional reputations, to account for matters that cannot be objectively measured. For example, calculation of GNP and GDP to measure a country's economic growth without taking into consideration environmental factors such as erosion, water pollution, disappearance of forest land and depletion of other non-renewable resources is nothing less than misleading the public. But this is exactly how GNP and GDP are calculated.

Democracy, Economic Freedom, Corporate Democracy and Capitalism

It is not the purpose of this book to discuss democracy, on the other hand, there is a misconception that once a nation practises a certain brand of democracy, it will facilitate capital accumulation and stimulate entrepreneurship to provide opportunities for individuals "to make money", leading to a paradise of economic freedom. Convenient as it may be, unfortunately, it is far from reality. This makes it necessary to clarify matters before proceeding with the exploration of entrepreneurship and corporate management.

The simplest way to understand democracy is to appreciate how it works in countries that promote democracy, for example, the United States, a country made up of people with diverse religions and backgrounds, but working together as a nation. Although not everyone agrees that the US should be viewed as a model nation of "democracy", there are people who like to think and insist that only the US version of democracy is the model for the world. The open-minded individuals, on the other hand, easily appreciate that while practising democracy is like countless motor vehicles travelling on the same road, it is virtually impossible to have every vehicle go through the same track. "Democracy" to Americans means:

1. Everyone is entitled to have his or her "private perception" about what his or her country is.

2. Every individual has his or her own right to express his or her views and opinions; at the same time, they must not prevent others from having and expressing their opinions and views.

3. Any individual has the right to be right and the right to be wrong.

4. All individuals are protected by the constitution.

5. Every individual has the right to be defended that he or she is not wrong.

6. Democracy is not without the commitment to the common good.

Under these circumstances, Americans are bound together to work for the common good. This, in fact, is the democracy we believe in, but only in "politics", not in the economy. Particularly, the economy of capitalism as currently claimed by politicians and academics, because capitalism as we know it today contains no element of "common good" which is the essence of democracy, rather, it is a system of "money talks".

For some strange reason, people in the high echelons of Western nations tend to perceive a direct relationship between "political democracy" and economic freedom, and infer that a country which enjoys a high degree of political democracy will promote growth. Unfortunately, facts do not support this belief. For example, Korea, Singapore and Taiwan, the first generation of the newly industrialized countries, and China, Indonesia and Chile, the second generation of industrialized nations, have quite different levels of "democracy" than Western political leaders would like to believe. In fact, Robert Barro, a *Wall Street Journal* contributing editor, would consider these "non-democratic" countries (see the *Sunday Times*, Singapore, December 1994, Focus section, p. 13); yet they have achieved high economic growth, and high standards of living – measured by real per capita GDP, life expectancy and literacy. It is true that these countries tend to become more democratic over time. For example, Singapore is projected to increase its democracy index from .33 in 1993 to .61 by the year 2000. Similarly, the projected value in the year 2000 for Indonesia is .43, whilst for Algeria it is .33 and for Syria .32. The last two countries had a .00 democracy index in 1993. Nevertheless, the effect of more democracy on growth is ambiguous. Negative effects are promoted by a tendency towards redistribution of income to the poor under majority voting and the enhanced role of lobbyists and interest groups in systems with representative legislatures.

In recent years, there has been wide acclaim for the economic miracles of ASEAN (Association of Southeast Asian Nations) and other Asian countries. Just how their democratic inclinations fit certain models is of no material significance; what does matter is that economic growth can happen in any political system so long as there are:

1. Responsible governments, with political leaders who truly provide leadership.

2. Policymakers who are willing to venture into creative and innovative schemes that will stimulate economic growth, avoid unworkable policies and do not repeat failed policies.

3. Supportive corporate citizens and people in general.

4. Large corporations and small to medium sized enterprises working as partners for the nation's economic growth.

In short, political democracy does not automatically provide economic freedom because the dominant force of large corporate power interferes with the processes of labour mobility, vibrant competition, unlimited entry, open information and consumer sovereignty, making it difficult for less fortunate people to gain access to the resources that lead them to complete "economic freedom". This is the goal of a responsible government.

Corporations are vital to a nation's economic growth, the well-being of the global economy and the well-being of humanity as a whole. On the other hand, not all corporations are necessarily created and developed in the best interests of society. The process of accumulation of private capital is often purely in the interest of financial gains for certain individuals or groups of individuals.

A simple observation of how "democracy" is practised among "democratic" nations will make us realize that what we claim is "democracy" is, in effect, a political system modified to fit the individual nation's needs. Be it the need of the people as a whole or a few individuals, it is yet to be defined. But we do have countries like the US, Canada, the UK, Australia, Switzerland and Costa Rica, as well as countries making progress in human rights such as Mexico, Haiti, Vietnam, Russia and African nations like Mali, Zambia and Niger. It is not difficult to appreciate the differences in political democracy these countries represent. What is less clear is the varied impact their policies have on economic development and growth, economic freedom and economic democracy. Complicating matters, most people only recognize their own brand of democracy and do not realize that true democracy does not exist, since absolute political democracy must be based on complete economic freedom. The relationship between political democracy, economic freedom and economic democracy may be expressed as:

- Absolute political democracy = complete economic freedom = economic democracy
- Greater political freedom = greater economic freedom, but not necessarily greater than economic democracy
- Economic democracy = business (or property) ownership widely spread among participants = entrepreneurship
- Entrepreneurship is defined as a wealth creation and value adding process in the interest of the individual and society (see formal definition in Kao, 1995)
- Entrepreneurship = common good
- Entrepreneurship is a function of (political democracy = economic freedom = economic democracy) = (business [property] ownership [power base, decision making unit] + management + working population)

In addition to its close relationship with political democracy, economic freedom and economic democracy as noted above, entrepreneurship is also a body of knowledge that consists, at least, of five subsets that the author could identify:

1. Creative innovative individual: entrepreneur (includes venture founder and owner/manager of his or her own business and

continues to be creative and innovative in all his or her economic endeavours).

2. Government entrepreneurship: public service entrepreneur.
3. Institutional entrepreneurship: such as entrepreneurial teacher.
4. Corporate entrepreneurship: corporate entrepreneurial manager.
5. Corporate entrepreneurship: corporate employees in non-managerial positions who are stakeholders in the entity. (As stakeholders, they make "on job" proprietary decisions. They see themselves as job or position owners and not as employees.)

Is There Such a Thing as Corporate Democracy?

Corporate democracy is a term used among academics to mean that democracy exists in the corporation. The model essentially refers to employee participation in management.[1] There are two groups of practices: internal decision-making participation and external decision-making participation (Smith, 1978).

In the case of external participation, there are two types of board structure:

- Popularly known as a European practice where employees are allowed to participate at the board of directors' meeting, while board members largely represent shareholders, and, particularly, block shareholders.

- Commonly acknowledged as an American style where one or several board seats are provided for individuals to serve as a voting member or members of the board. The root of the challenge is the accounting bias of treating "human resources" as an expense rather than an asset. Since it is an expense, it is therefore expendable as it is incurred. It is never in the mind of managers or, particularly, accountants that human resources should be considered as a corporation's valuable assets.

Why this Book?

This book is not perceived as a book of "corporate management" that follows the conventional approach of discussing management skills and technologies, nor is it the traditional approach that describes matters such as leading, planning and controlling an organization, and the more exciting ideas of management of conflict and change, strategies, management of information, the use of war games to theorize management practice or competition.[2] Efforts will be made to put business in the context of philosophy and corporate purpose, with a broad approach of understanding the nature of profit, sensible resources management, and environmental issues. It will be based on the social entity concept that requires managers to assume responsibilities for all contributors and humanity through entrepreneurship, creating wealth for themselves and adding value to society.

The beginning of the beginnings

To simplify the argument, the beginning of all corporate management challenges can be simply referring to the question of "making a profit". It may sound too simple to believe that in the mind of a corporate manager, making a profit is the first, last and always the purpose of their "professional career". Corporate managers seldom attempt to explore the nature of profit, and what profit does to themselves and others. As a matter of convenience and trust, they tend to leave the issue of "profit" with accountants, and accountants tend to rely on economists for guidance, but corporate managers with their own ways of thinking will do what they think is best to safeguard their corporations' interest. Any irregularities will either be considered of no "material significance" or "let's do more study, one study after another but no steps taken for fear of distorting that which will risk professional reputation". In short, no fundamental changes are necessary.

Basically, it involved three different but related irregularities that have led us to think that we are practising corporate management under what people are fondly referring to as the "market economy" and invisible hands pricing system. The matter relates to profit and value, since there is no apparent attempt made to discover what is value added and what is profit, the other concerns the professional responsibility to the erosion of environmental health and draining of natural resources, and the third is the damage of micro-theory that is employed in education and business practice.

1. *Economic endeavours can only increase utility functions through making different combination, there is really no such thing as "profit".* The reality is that if we take a broad approach there is really no such thing as profit, if we take humanity and planet earth as a whole, because we cannot add anything to the earth, all we have is to take what the earth and the environment make available to us, and we make different combinations to increase the utility function for human consumption and for other living creatures as well. "Profit", if put bluntly, is either taking from people who contributed to the process or from the resources pool and environment which supposedly made available to us now and in the future. If "profit" is not recycled back to the people, the resource pool and the environment, we will have even more serious troubles ahead of us by creating further departure of "economic democracy" from mankind or impairing the life support system of human needs. The challenge of corporate managers is not merely to make profit for today, but to add value to society.

2. *The professional responsibility to the mismanagement of the natural environment and resources.* The author worked as an international consultant to a UNIDO (United Nations Industrial Development Organization) enterprise development project in China last year. While travelling in the interior of China, through small towns and villages he noticed a few road signs which read: "It is our responsibility to develop our economy, it is the government's responsibility to deal with the environment." Through the efforts of environmentalists, the government and the private sector, some progress has been made to reduce the erosion of our natural environment and contamination of the environment's health; for example, new technologies have been developed for the logging industries (rather than cutting down old trees, branches and barks of young trees are used for industrial purposes) and chemical products that lessen the pollution effect on the environment have been produced. But they fail to understand the professional responsibilities to report natural and human sacrifices as part of production cost, and mislead the public by using absolutely "damaging" indices expressed in GNP and GDP as a nation's economic growth. Consequently they lead people to believe that "profit" is a gift from "heaven", and never give a thought to the fact that it is really resources and the environment which sustain the lives of future humanity and other beings. Is it possible to expect the renewal of the "Wealth of Nations" and continuation of life, if we only think of and pursue

the maximization of the rate of return on investment (capital), without the return on the environment and resources?

3. *Micro-theory that killed the need for common good of economic endeavours in cold blood.* Anyone who has had a business or economic education must have had some exposure to the classic work *The Wealth of Nations,* written less than 200 years ago by Adam Smith. Even without going into any great detail about the book, we can still appreciate the essence of Smith's philosophy. The "common good" is expressed throughout the deliberation, and it has a profound effect on the thinking of people not just in the English-speaking universe, but people with diverse culture as well. Unfortunately, microeconomists, perhaps for the convenience of economic analysis, began with their attempt to build mathematical models to advocate the theory of firm and profit, and divided economic discipline into two parts – the macro and micro, as they are unrelated in making economic progress for mankind. While macroeconomics supposedly has to do with the "common good", microeconomics turned out to be a monster working only for the corporate interest, and making profit, without troubling to build models and translate these models into working terms to deal with the "real cost" involved in earning corporate profit. Until this day, there has been no real clear evidence that micro-economists made a serious attempt to close the macro–micro gap. Opportunists see the chance to express their disenchantment with government intervention by advocating "free enterprise", even in cases of common good. The consequences are not difficult to note; those who are rich become richer, and those who are poor, try anything to make themselves rich. If you cannot make yourself rich under the "free enterprise" economy, too bad.

The author had an opportunity to discuss "free market economy, pricing system being the invisible hands," with an economics professor. He said that the pricing system will make all the adjustments necessary. For example, water used to be free, but if we want pure water to drink, we have to pay a price for it, the more the demand the higher the price will be. High price will stimulate entrepreneurs to find new sources of water because they are lured by profit. The author asked the economics professor: "would we have to buy air one day?" He said "yes", the cleaner the air you want, the higher the price you will have to pay. And the author asked him, if you do not have the money to pay for the necessary air to support your life, would it be true to say that you will be dead. He responded without hesitation, "You are right, Raymond, you will be dead. This is why we are exploring the possibility of migrating to another planet." Oh! If we rely on the market economy and the pricing system and successfully explore the unknown before we can have air to breathe, how many people will be dead? Then, this will also be part of the market economy, because it creates capitalistic business opportunities to dispose massive death because of the shortage of clean air!

It is because of the above reasons that the author was encouraged to write this book, with a simple purpose: corporate managers can desire to make all the profit they desire, but they must know what the profit is and how it works and affects people. The shareholders and top executives may take all the corporate earnings they want, except they must know where the "profit" came from. Is it from financial income generated from the accounting process, or is it operating residual from a surplus after distribution of legitimate cost under the present circumstances, and/or from exploiting the environment and life support of our future generations? The book's purpose is not to illustrate how a corporation should be managed to make profit, but to add value to society and create wealth for the individual in the process. Harming others and the environment may create financial wealth for the individual, but it does not add value to society.

About this Book

Since the informal ending of the Cold War and the disintegration of the Warsaw Pact, countries which previously were practising command economy are all now turning to the other extreme – the practice of capitalistic economy. They hope that through rapid capital accumulation, "entrepreneurship" will be stimulated; through "entrepreneurship", they hope to create a generation of rich people (some individuals in the academic community label the process as "the dynamics of capitalism"). Seldom do they trouble themselves to question how such a transformation is to be made, and what "rich" means to the individual or the society?

In a broad sense, the relationship between economic growth, capitalism and entrepreneurship may be expressed as shown in Table 1.4.

The illustration suggests that the whole idea of capitalism is people's desire to invest purely for financial gains. However, investment can stimulate entrepreneurial activities as well as business activities. The ideal situation is entrepreneurial business undertakings inducing investment, which in turn, will stimulate more entrepreneurial activities. This book attempts to explore the opportunities to affect the corporate managers' mindset and encourage the development of entrepreneurial skills that will attract resources to engage in entrepreneurial business activities, making corporations not just money-making dream machines for a few, but communities of entrepreneurs determined to create wealth and add value to society through the effective and meaningful utilization of limited resources in the interest of humanity.

The approach of this book initially is intended to be thought-provoking. Therefore, the first section, consisting of six chapters, will deal exclusively with issues of controversy. The introductory chapter provides a closer examination of corporations' function in society and matters related to political democracy, economic issues and entrepreneurship. It includes the issue of profit, management theory, the general management model followed by business schools throughout the world, and how they impact managerial decisions. In addition, it considers environmental issues and the cost of capitalism, which we tend to take for granted without really exploring the nature of it, and how it affects our lives today and in the future. More importantly, corporate managers are encouraged to recognize their challenges, not just for profit or meeting the ROI requirement, but to manage resources and serve human interests at the same time. The subsequent deliberation will lead to an examination of the necessity of adopting entrepreneurism as an alternative to capitalism in the one extreme and communism on the other. This is followed with the application of entrepreneurism and entrepreneurial strategies to renovate the organization through the use of broad application of the stakeholder concept to prompt individuals in the corporation to make entrepreneurial decisions. The "tools" managers used for decision-

Table 1.4 Relationship between capitalistic expansion, entrepreneurship and perceived conventional economic growth

Player	Player's desire, need and/or want	Action	Result
Entrepreneurship	Those who have the inner need to do something new and different	Creativity and innovative entrepreneurial undertakings	Creating wealth for the individual and adding value to society Return the residual and facilitating further capital accumulation
Capitalistic expansion	Those who have capital and the desire to invest for the purpose of rapid capital accumulation	Investing in money-making schemes of normal business undertakings	Increasing financial wealth and further capital accumulation Return the profit and facilitating further capital accumulation
Perceived conventional economic growth	Those who have the inner need to be rich and for rapid capital accumulation of personal financial wealth	Engaging or investing in business undertakings regardless of social consequences or the effect of others' interest (with or without knowledge of criminal activities that will harm others and the environment)	Increasing personal financial wealth but adding no value to society or causing harm to society and environment

making are then examined. The introduction of entrepreneurial management focus and process as well as the transformation of the current practice make a corporation a community of entrepreneurs for the purpose of creating wealth for the individual and adding value to society. The deliberation may be expressed as shown in Table 1.5.

Table 1.5 About this book

Objective	Section highlight	Chapters
To recognize and understand that a corporation is not a money-making dream machine, but a community of entrepreneurs, created for the common good and the creation of wealth for the individual	Introduction and background. Introducing entrepreneurism as the alternative to communism and capitalism	The corporation and its relations with market economy, political ideology, entrepreneurship and entrepreneurism (Chapters 1–6)
	Entrepreneurial value as the basis of corporate structure	The Unifying Theory, vision, corporate culture, entrepreneurial value (Chapters 7–9)
	A corporation as a community of entrepreneurs	Corporate value and enterprising culture (Chapters 10–11)
	Entrepreneurial environment	Mindset, entrepreneurial attributes, development and organizational strategy and staff function in an entrepreneurial environment (Chapters 12–13)

Table 1.5   (cont'd)

Objective	Section highlight	Chapters
	Entrepreneurial managers	Resources constraints, management and environmental considerations in corporate strategy (Chapters 14–15)
	Corporate reporting	The value added approach to corporate reporting (Chapter 16)

Definitions

For the purpose of this book, the following definitions are used (Kao, 1995, p. 84):

- **Entrepreneurship** – Entrepreneurship is the process of doing something new (creative) and/or something different (innovative) for the purpose of creating wealth for the individual and adding value to society.

- **Entrepreneur** – An entrepreneur is a person who undertakes a wealth-creating and value-adding process, through developing ideas, assembling resources and making things happen.

- **Enterprising culture** – Enterprising culture is a commitment of the individual to the continuing pursuit of opportunities and developing an entrepreneurial endeavour to its growth potential for the purpose of creating wealth for the individual and adding value to society.

Please note, these broad definitions suggest that every individual is an entrepreneur at some point, but no individual is an entrepreneur all the time because the essence of the definitions places emphasis on the need for "creativity", "innovativeness", "wealth-creating" and "value-adding" in practice.

The Impact of Corporate Governance on Corporate Entrepreneurship

TAN WEE LIANG

Nanyang Business School, Nanyang Technological University, Singapore

Introduction

As the twenty-first century approaches, two areas will be significant factors in the continued success of companies and their contribution to global economic well-being. The first area is corporate governance. The corporation of the twenty-first century not only needs to consider the interests of their shareholders, but also the demands of government and social and environmental concerns. Corporate governance also plays an important role in economic development. Researchers have recently examined whether the traditional model of corporate governance would be ideal for emerging economies as there was evidence to show that a greater role for institutional members such as banks might be more appropriate (Aoki and Kim, 1995). Other researchers are re-examining corporate governance with the view of recommending appropriate systems of governance for the twenty-first century corporation (Blair, 1995).

At the same time, companies need to introduce corporate entrepreneurship. This is the second area of importance. Companies have recently discovered the need to innovate internally. Many companies have sought to encourage their employees to innovate and be creative. Celebrated companies such as 3M in the US are commended for their success in inventions and new ventures started internally. Many other companies aspire to corporate entrepreneurship, also known as intrapreneurship (Pinchot, 1985). The reasons are fairly obvious. In order for companies to be competitive, they need to be innovative in their product or service offerings, their processes and also in management. Small entrepreneurial firms have shown themselves to be flexible and highly innovative. Larger companies are interested in transforming themselves so as to incorporate the entrepreneurial spirit in their organizations.

This appendix explores the relationship between the two areas, corporate governance and corporate entrepreneurship. It is suggested that the two are interrelated and that any efforts to introduce corporate entrepreneurship are affected by the existing systems of corporate governance.

The Impact of Corporate Governance on Corporate Entrepreneurship

Initially, the two topics may appear to be unrelated. As it is traditionally understood, corporate governance has become part of the boardroom vocabulary (Kendall, 1994) and refers to questions about the structure and the functioning of a boardroom of directors or the rights and prerogatives of shareholders in boardroom decision-making (Blair, 1995). Disclosure and control over management's activities would be areas of importance under this narrow view. The guidelines in the Cadbury Code of Best Practice are examples of the elements in such system of corporate governance. The traditional model is depicted in Figure A.1. In this model the company should serve the interests of the shareholders since they are the ones with the "residual" claim (i.e. they are the ones who are paid last in the event of a winding-up and in the usual course of events after the suppliers, the employees and the creditors). The interests of the management and the employees are secondary to those of the shareholders; the owner should have control. It has also been referred to as the doctrine of shareholder sovereignty.

However, corporate governance is capable of a wider definition. Under the wide view corporate governance encompasses the company and all the interested parties, including the traditional insiders and, in addition, the outsiders: suppliers, employees and lenders. It includes the whole set of legal, cultural and institutional arrangements that determine what companies can do, who controls them, how that controls them, how that control is exercised and how the risk and returns from the activities they undertake

Figure A.1   The traditional model of corporate governance

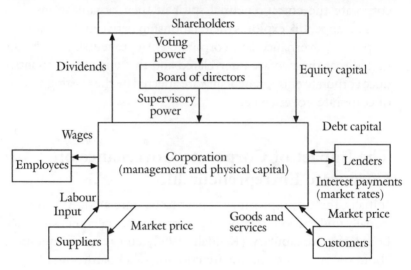

Source: Blair (1995).

are allocated. The wide view of corporate governance has been called the stakeholder approach (Alkhafaji, 1989). The stakeholders are those people who have an interest in the company's activities including members of the external environment such as the community, unions and consumers and those of the internal environment such as the management and the stockholders. This concept of corporate governance is multidisciplinary in approach, encroaching upon areas such as corporate law, organization theory, economies and management. It suggests that the companies should serve the interests of the various stakeholders. Corporate governance would include all the systems of governance in a company including human resource accountability and reporting, and organizational structures.

Entrepreneurship, on the other hand, is often associated with new venture creation, innovation and creativity. Entrepreneurs are now recognized as contributing to the economy through the jobs they create (Birch, 1982), the new combinations they bring about, and the economic "catching-up" of certain economies such as Hong Kong (Cheah and Yu, 1995). Corporate entrepreneurship refers to the efforts on the part of companies to foster the spirit of entrepreneurs, innovations and new ventures in the corporate setting. The process usually only involves selected employees – technicians, engineers and certain managers. It may be confined to selected units in the organization entrusted with selected ventures or ideas to experiment with. However, corporate entrepreneurship, like corporate governance is capable of a wider definition. Kao (1995) refers to a state of affairs where every employee is motivated to be involved in creating wealth and adding value to himself and to the company. By his definition, an entrepreneur is a person who is involved in the process of doing something new or something different to add value and create wealth for himself and for society. This definition is an inclusive one. It calls for any individual whether in a profit or not-for-profit organization to be involved in entrepreneurship.

Although corporate governance and corporate entrepreneurship do not appear to be related, there a number of ways in which corporate governance in the narrow and wide sense impinge on any efforts to encourage entrepreneurship. Corporate governance under the wide view would encompass any mechanisms put into the company to foster internal corporate entrepreneurship. The effects of corporate governance on corporate entrepreneurship are clearer when you examine the ways in which companies have sought to encourage internal corporate entrepreneurship.

Schollhammer classified (1982) the various ways in which corporate entrepreneurship have been engineered in companies in the following manner:

> *Administrative entrepreneurship.* The company engaging in this form of corporate entrepreneurship is taking a step beyond having a traditional R&D department. There is a philosophy of corporate enthusiasm for supporting researchers, accompanied by the provision of extensive resources for making the new ideas commercial realities.

> *Opportunistic entrepreneurship.* The company encourages champions to pursue opportunities for the company, and through external markets. The example given is Quad/Graphics, the company that prints *U.S. News & World Report, Inc.*, and *Newsweek* among others, permitting its managers to develop technology for printing with computers and even selling the technology openly to anyone (Holt, 1992).

Acquisitive entrepreneurship. Under this category the corporate managers are permitted to seek new opportunities outside the company. These may include mergers, acquisitions, new technologies and strategic alliances.

Imitative entrepreneurship. This form of entrepreneurship is epitomized by Japanese firms in their study and reverse engineering of others' products.

Incubative entrepreneurship. Semi-autonomous new venture development units are formed that provide seed capital, access to corporate resources, freedom of independent action and responsibility for implementation from venture concept to commercialization. The picture here is one of in-house incubators.

The list is not exhaustive as it does not include Kao's category of entrepreneurs present in a corporation. Be that as it may, you discover that underlying each of the classifications is the implication that there is a supportive environment in the company and that endeavours benefiting both the innovators and the company are welcomed. It can also be seen that under the wide view of corporate governance, the system of governance could either help or hinder internal corporate ventures. Thus, in designing the system of governance over human resources and departments, it may be necessary to permit the existence of special units to spearhead internal corporate projects. One example of such a unit that is becoming a usual feature in most corporate structures is the Research and Development (R&D) Department. A few companies employ the spin-off approach (incubative entrepreneurship) where the employees are encouraged to spearhead projects for which they are given autonomy but financed by the parent company. They may be permitted to purchase shares from the parent company subsequently. Variations on this theme could exist where the parent company permits the "skunk works" to operate on a special budget under one of the regular divisions of the company. In some companies, these units are perched under the most appropriate division.

At times, instead of formal arrangements there are informal groupings of staff. Hence, in designing governance systems over the employees that are ideal, those responsible have to be mindful that instead of ideal and clear-cut systems, there may be a need for loose groupings of staff across functions if internal projects are to be encouraged. In 3M, for example (Peters and Waterman, 1981), employees are encouraged to pool resources for any projects that they may wish to pursue. Innovation champions are encouraged. Such arrangements may run contrary to neat lines of communication and accountability that are traditionally prescribed. Neat job descriptions and performance appraisals based on those descriptions would not encourage cross-functional relations within companies. Yet it is cross-functional teams that often develop products that perform well. It is not the team of engineers who design the product but a team that may encompass engineering, marketing, design and customer service quality personnel. The Japanese system of corporate governance has been commented on and has been of interest to the West precisely because there is an emphasis on teamwork (Peck, 1988).

The narrow view of corporate governance would not be concerned with the measures that encourage corporate entrepreneurship highlighted above since it confines corporate governance to the boardroom, only. Yet there is another threat to corporate entrepreneurship which is veiled in both views of governance. This veiled threat is addressed in the next section.

Corporate Governance, Profit and Corporate Entrepreneurship

Present day corporate governance, whether narrow or wide, has the underlying objective to facilitate the mission of the company: profit. The traditional model of corporate governance seeks to protect the interests of the shareholders. The investor-shareholders' primary concern is profit. Profit is still a major consideration even under the wide view of corporate governance. The mission of the company has not changed, despite the call to consider interests of the other stakeholders under the wide view of corporate governance. The mission of the business, its raison d'être, is still profit. The wide view of corporate governance calls attention to corporate social responsibility, ethics, government and the environment. However, the systems of governance are still intended to facilitate the protection of shareholders' interests. Few have argued that the company should abdicate its prerogative to decide on the basis of profit and permit the other stakeholders' interests to prevail. Even if the other stakeholders' views were given prominence, most decision-makers would "vote" with the continued prosperity of the company in mind.

The profit motive and the corporate governance mechanisms put in place would run counter to the efforts to encourage corporate entrepreneurship. Kao (1995) argues that if profit continues to be the mission of the corporation and if the rate of return on investment (ROI) continues to be the sole means by which the performance of a company or its units are measured, corporate entrepreneurship has failed its blood test. His reasoning is simple. Projects that deserve to be championed may be cast aside as non-starters because of the financial performance parameters. If the profit centre concept is applied to the units pursuing internal corporate ventures, the units may be shut in a short period of time for the corporate governance system would peg survival to the continued profitability of the units concerned. Granted that there may be a grace period, the day of accounting would still arise. The managers responsible for these units would be mindful of their performance criteria as well. Theirs would be an existence dependent on the success of projects that they pick. It would be reasonable to presume that projects that engender results in the short-run would be selected and not those projects where the returns may be beyond a five-year time frame. Yet great value might accrue from the rejected projects. It would appear to be a paradox as profit will be generated when the corporate entrepreneurial ventures are successful. Yet when profit alone is the criterion for selecting or initiating projects, it can prove to be the death of proposed projects.

The objective that is served also serves as a motivator of behaviour on the part of employees. Performance criteria pegged to profit have a signalling effect on workers, particularly when the rewards are not shared with them. The budgeting process engaged and the quarterly returns that have to be met send a particular message to the employees – not necessarily the ideal one that says that their efforts are and will be appreciated but more likely that "the bosses so require".

A challenge thus confronts those who design corporate governance systems in the company. The organization's hierarchy and structure, lines of communication, performance appraisal criteria and culture are part of the corporate domain. The corporate governance systems should serve to motivate staff and reward intrapreneurship. They have an impact on corporate entrepreneurship.

Conclusion

Corporate governance is important as it forms the environment for the internal activities of a company. Although in the traditional model, it is confined to the composition, selection and powers of the board of directors, it should not be confined to those concerns. The multidisciplinary nature of corporate governance can be seen in the consideration of corporate entrepreneurship. Corporate administrators and custodians of governance ought to realize that governance needs to be in line to facilitate other corporate objectives.

The company of the twenty-first century will need to evolve in line with the needs of the world and the corporate concerns. A balance needs to be struck between those concerns. What is clear is the need for corporate governance to accommodate the need for the company to have intrapreneurs. Corporate entrepreneurship will not flourish unless the environment is appropriate.

2

Beyond just word play, is there such a thing as "profit"?

If the capital of a business is impaired, there is no profit.

If underground mineral deposits are diminished, there is no profit.

If unrenewable resources are expended, there is no profit.

If fishermen take the last spawning fish, there is no profit.

If virgin forest is destroyed for cattle grazing or for industrial use, there is no profit.

If a country makes a living trading only on the basis of resources, there is no profit.

If a country has a high rate of unemployment and massive poverty, there is no profit.

If management is lagging behind technology, there is no profit.

Unfortunately, we know a lot about making money, but so little about "profit", and do not realise that profit does not come above people and environment.

The word "profit" is used by corporate managers, as if it is the only thing that matters in business. Admittedly, there is no doubt that without profit, no one will be interested in investment, a corporation could not sustain itself in the marketplace, and sure enough, without profit, there would be no growth possibilities. But the sad reality is that, whilst corporate managers, like others talk so much about profit, they seldom question the nature of profit, nor care about the consequences for people and the environment in the process of making profit. This chapter intends to highlight some crucial issues involved in making profit. The objective is twofold: (1) To increase the awareness of corporate managers to appreciate what real profit is; and (2) To indicate how to determine corporate profit, and be prepared to accept the responsibility of making profit on the one hand, and have concern for people, resources and the environment on the other.

Is There Such a Thing as Profit?

Judy, a humanities student, posed a question on the topic of making profit to her business and academic friends at a small gathering. Questions related to profit, business or humanity stimulate broad discussions. After a few rounds of arguments, Judy showed the group a newspaper clipping (from *The Straits Times*, Singapore, 30 July 1995, Life at Large section, p. 9), bearing a Pulitzer Prize photo showing an emaciated girl collapsing on the way to a feeding centre, as a plump vulture lurked in the background. The caption reads as follows: "Kelvin Carter, the South African photographer whose image of a starving Sudanese toddler stalked by a vulture won him a Pulitzer Prize this year, was found dead on Wednesday night." Judy added: "He chased away the vulture, and sat under a tree for a long time, smoking cigarettes and crying after witnessing the Sudanese girl collapsed. Police said that he had apparently committed suicide. He was 33."

Judy then asked the group: "If there is such a thing as profit, where does the profit go and what does it really mean to humanity when some are making profit for themselves while others are starving to death and ending up as vultures' feast?"

Peter intervened: "Judy, I don't know why you raise this question, what does profit have to do with Carter's picture? Profit means money gain, or excess of return over outlay (based on the *Oxford Dictionary*). It is a measurement of business performance, or individual transaction in the market economy. It does not mean that when a firm makes profit, people will starve, or that a vulture will feast on human flesh."

The discussion went further into the distinction between profit for the individual or an individual firm to GNP, GDP and a nation's economic growth. But the group was puzzled by what really causes tragedy. Does "profit-making" really contribute to the economic good of all humanity or does it divide people into the rich and the poor? On the other hand, taking humanity as a whole, is there such a thing as profit? And if so, where does profit come from? From hard work? From an individual's intelligence? From other people or from the environment? At this point, the group realized that the discussion had been carried far beyond its original intent, and fearing the never-ending arguments of ideologies, the discussion ended without satisfaction.

Profit, as We know It

To most of us, profit is as simple as what Peter quoted from the *Oxford Dictionary*. Alternatively, the *Webster Dictionary* states: "profit is the excess of returns over expenditure in a transaction or series of transactions." As those definitions are simple enough to comprehend, we tend to take them for granted. We seldom question their relevance, origin and how they work or affect people's lives. On the other hand, defining profit is for accountants and economists, a really soul-searching exercise. However, while economists tend not to bother too much with reality, accountants take reality into their own hands, and impose, with the help and hindrance of various collaborations and pressures imposed by the stock exchange commissions, their concept of "profit". As a result, measuring, interpreting and using it dominates the business world as we know today. If the question, "what is profit?" is raised, do not be surprised if it provokes a reaction such as: "Profit is profit, who cares who says what!"

Accounting Profit or Accounting Income

In the field of accounting research and practice, the earliest authoritative work on profit determination can be found in the work of Paton and Littleton (1940). It established that profit was based on the separation of investment and management, which is derived from the "Theory of the Firm" advanced by economists some 100 years ago, and followed by a group of concepts, including:

- Entity
- Continuity of activity
- Measured consideration
- Cost attached
- Efforts and accomplishment
- Verifiability
- Objective evidence
- Assumptions

Since then, a great deal of effort has been made to solidify the concepts of cost determination, revenue, income and surplus (Paton and Littleton, 1940, pp. 7–23). In simple terms, all the above form a conceptual base that provide a needed justification for accounting practice, in particular, corporate profit determination.

Some twenty-five years later, Paul Grady (1965) completed his *Inventory of Generally Accepted Accounting Principles for Business Enterprises*. Known as Accounting Research Study No. 7, published by the American Institute of Certified Accountants Inc., New York, it gave direct reference to the stock exchange in dealing with matters of accounting practice. As noted, the New York Stock Exchange in the early 1960s participated with the Institute committee in laying the foundation for generally accepted accounting principles which are in fact based on a series of postulates originated from ten concepts much the same as described by Paton and Littleton as follows (Grady, 1965, p. 24):

1. A society and government structure honouring private property rights.
2. Specific business entity.
3. Going concern.
4. Monetary expression in accounts.
5. Consistency between periods for the same entity.
6. Diversity in accounting among independent entities.
7. Conservatism.
8. Dependability of data through internal control.

9. Materiality.

10. Timeliness in financial reporting requiring estimates.

On the basis of these concepts, a series of postulates were developed by combining them with an early work of the American Institute of Certified Public Accountants (AICPA), known as ARS (Accounting Research Study) No. 2, which included postulates governing financial statement, market price, stable units and disclosure. These led to the formulation of three sections of Generally Accepted Accounting Principles that govern accounting practice, which include:

- Principles A1 to A9 designed to measure revenue, costs and income determination

- Principles B1 to B7 attempted to measure equity capital invested by shareholders through contributions of assets or retained earnings

- Principles C1 and C2 prepared to measure the assets invested by stockholders (through property contributed or retained earnings) and creditors

After the publication of Grady's *Inventory of Generally Accepted Accounting Principles for Business Enterprises* (1965), the American Accounting Association published *A Statement of Basic Accounting Theory* (1966). This was, perhaps, the first time that the accounting professionals in the US decided to formulate a theory which did not confine the accounting practice to areas of providing information, but extended its involvement and responsibilities to making decisions concerning the use of limited resources. These resources involved not just the material resources, but human resources as well as facilitating societal functions and controls with the following defined objectives:

1. Make decisions concerning the use of limited resources, identification of crucial decision areas, and determination of objectives and goals.

2. Effectively directing and controlling an organization's human and material resources.

3. Maintaining and reporting on the custodianship of resources.

4. Facilitating social functions and controls.

On the basis of these objectives, the statement specifies the scope of accounting, and methods of the current accounting practice and applies the standards to areas presently considered to be outside the accounting function. Suggestions are also made for practitioners to observe the following guidelines with respect to communication on accounting information rather than previously noted standards:

1. Appropriateness to expected use.

2. Disclosure of significant relationship.

3. Inclusion of environmental information.

4. Consistency of practices through time.

In short, the statement provides the following sequential development as shown in Figure 2.1.

Figure 2.1   From accounting theory to guidelines for accounting practice

---

Accounting theory

A cohesive set of hypothetical, conceptual and pragmatic principles of reference for a field of study (as defined by the Committee to Prepare a Statement of Basic Accounting Theory).

Objectives

1. Make decisions concerning the use of limited resources, including the identification of crucial decision areas and determination of objectives and goals.

2. Effectively direct, control and organize human and material resources.

3. Maintain and report on the custodianship of resources.

4. Facilitate social functions and controls.

Figure 2.1 (cont'd)

Suggested future scope

1. Objective of the accounting function: measurement and communication of data revealing past, present and prospective socio-economic activities.

2. Underlying disciplines used in performing the accounting functions: behavioural sciences, mathematics and less well-defined areas such as information theory and computer science, as well as the conventional accounting methods.

3. Purpose of accounting function: to improve control methods and decision-making at all levels of socio-economic activities.

Standards for accounting information

1. Relevance.

2. Verifiability.

3. Freedom from bias.

4. Quantifiability.

Guidelines for communication of accounting information

1. Appropriateness to expected use.

2. Disclosure of significant relationships.

3. Inclusion of environmental information.

4. Uniformity of practice within and among entities.

5. Consistency of practices through time.

It should be noted that while the American Accounting Association (AAA) is an organization consisting of a body of academics, the American Institute of Certified Public Accountants (AICPA), on the other hand, is a professional organization. Although there is cross membership, and they both intend to strive towards a common goal, the AAA tends to be interested in the pursuit of knowledge, and under this assumption, accounting has become an academic discipline, with:

1. Sophisticated methodologies for information processing including advanced computer and statistical analyses.

2. Affiliation with a variety of accounting entities, including national organizations.

3. A functional and theoretical separation of accounting activities representing selected points in the flow of socio-economic activities such as transactions and other objective points in the flow process.

The AICPA, on the other hand, is primarily interested in the practical aspects of the information process, and reporting in financial terms; more specifically, to satisfy stock exchange requirements.

Although it is almost three decades since the AAA published *A Statement of Basic Accounting Theory* to acknowledge the need for accounting to consider the broad issues of resources limitation and societal functions, the accounting practice has shown no clear attempt to recognize these realities. Since both resources limitation and social functions are part of the cost of doing business, failure to recognize these realities would understate the cost of doing business. Similarly, profit reported would not be the current profit alone, but inclusion of part of the capital which will be needed for the future. However, as the current accounting practice is primarily based on the concepts established in the Paton, Littleton and Grady days, these governing concepts are the ones which need to be questioned. To some extent, these concepts are interrelated such as monetary expression relating to conservatism. However, anything not readily measurable in monetary terms is unlikely to be recognized by the accounting practice. There are a few exceptions that are significant enough with respect to accounting measurement to be recognized, such as the utilization of limited resources, the determination of profit including matters of private property rights, bias in financial reporting, monetary expression in accounts, conservatism, materiality and timeliness in financial reporting that require estimates.

1. Private property rights

Recognition of private property rights is a fundamental requirement in a capitalistic society, and capitalism must be protected by the law. Therefore, property rights are part of the accounting practice. On the other hand, the utilization of capital assets in business activities does not merely involve capital assets acquired through exchange, but also the utilization of limited natural resources and life support systems which are also property rights. Many of these rights do not belong to the present, but to future generations and in fact should be protected by the law just as much as private property rights. But in the accounting process, no attempt so far has been made to account for these rights. Familiar advertisements such as: "We are bringing the future for your present enjoyment" are an open admission that some business operations are in fact deliberately ripping off property rights of our future generations.

2. Biased reporting

One of the standards for reporting accounting information is freedom from bias. In fact, when estimates are required, it is almost without doubt that there will be built-in bias in estimation, such as depreciation or capital cost allowance. The greatest bias exhibited in accounting practice is still its bias in favour of revenue, while the same is not applied to cost, especially, the cost to society in the course of business operation.

For example, in the Atlantic coast provinces, inshore fishermen traditionally make a living by using relatively simple fishing techniques, a small boat, and land-based processing. The operation may be "inefficient", but it has provided jobs and maintained a society which is basically self-sufficient in a stable cultural (the working culture) environment. Meanwhile the large corporations, with the help of the government, turn large fishing vessels into fishing plants. Rather than fishing responsibly to meet society's needs, massive overfishing has probably not only exhausted the fish stock, but also wiped out the livelihood of smaller inshore fishermen and workers in the fishing plants on land. Consequently, the costs of entering into a large-scale automated fishing operation involve at least the following:

- **Cost to the environment:** there is a tendency to overfish and drain fish stocks.

- **Cost of job loss:** many smaller fishermen apparently lost their place in the fishing industry and this job loss results in social disorder. To remedy the situation, government intervention is inevitable. The intervention involves either handing out welfare cheques to displaced fishermen or finding other ways of restructuring the Atlantic economy.

- **Cost of culture and societal loss:** the transformation from a simple fishing society into other forms of society would result in the disappearance of a community's culture.

As one accounting objective with respect to social functions is to facilitate the operation of an organized society for the welfare of all, why is it then the cost to society, and the impairment of cultural stability of a working community, is not reported as part of the cost of large-scale automated fishing operation?

While accounting practice would recognize the revenue made by the fishing vessels of large companies as the result of large-scale fishing and the automated fishing process, the cost to society would not be recognized in the cost accumulation. Similarly, in the real estate business, gain in value as the result of social contribution is recognized, but the loss of alternative use of land, and the construction damage effect will not be included in the cost of a real estate business undertaking. The understated costs of these operations would result in the overstatement of profit and if there is a loss incurred, loss will be understated as well.

3.   Monetary expression

As money is the only common denominator in life and in the business world, no one would deny the reality that everything must be reduced to dollars and cents, otherwise commercialization would not be possible. Barter is a value exchange system that could, in fact, function as an exchange economy and is relatively common in Canada and the US. Under our market economy, however, money is still the only one that is recognized as the medium of exchange. A market economy without money is just unthinkable. The issue is not a question of whether to use money as the medium for exchange, but the way in which accounting practices measure costs. They favour only costs incurred that can be readily measured in dollars and cents. As such, opportunity costs are a form of sacrifice that must be accounted for in the cause of cost recognition, and ultimately profit determination. If such costs are not recognized and accounted for, profit would be overstated as a result.

Opportunity cost has been recognized in the legal system and is widely used. A person who suffers serious injuries in a car accident would probably be awarded thousands of dollars. Yet in accounting practice, this form of personal sacrifice is not acknowledged, largely because it cannot be readily expressed in monetary terms. Nevertheless it is an additional cost of doing business. While on the topic of opportunity cost, it should be noted that in some more aggressive accounting practices, the capital cost (a form of opportunity cost) is recognized. This is the idea seen earlier in Alfred Marshall's work, *Principle of Economics* (1920, p. 62).

4. Conservatism and materiality

Conservatism is closely related to materiality; in the course of communicating information, accounting only recognizes those transactions that can be materially established. Through computerization, the process of information retrieval on reporting may have been substantially improved, but the fundamental premise of information processing is much the same. Consequently, profit determination is still confined within what the practitioners perceive to be relevant to compliance with the accounting standards of reporting.

5. Timeliness

Economists are troubled with the inability to determine production periods. On the other hand, accountants are adopting the timeliness principle and by-passing the need to determine production period. The reliability of reported accounting profit very much depends on how a time period is determined and most importantly, how it coincides with the end of the production period. Under the circumstances, this is next to impossible. As such, real profit can only be determined if the business terminates its operation.

The Nature of Accounting Profit

On the basis of the above analysis, accounting profit, in its real form is a manipulation based on a set of principles derived from a set of postulates based on accountants' perception. Although it meets the requirements of the stock exchange and, to a lesser extent, managerial use, it is far from being profit representing a real gain for economic undertakings. It is no wonder that in the case of the Olympia Corporation, a billion-dollar business empire whose reported profits impressed bankers and many other investors, could crumble in a matter of days. There are also situations, where a business entity repeatedly reports not making any profit, but after a period without additional owners' investment, the size of operation expands manyfold. If profit means a real gain for economic undertakings how could these irregularities happen in the business world? On 15 November 1994 on the 18:00 CTV news, Toronto, Canada, an opposition member in the Ontario Legislature shouted out to the Minister of Finance of Ontario: "You have a discrepancy in the deficit of $8+ billion forecast by $2 billion." The Minister responded: "I am using a different method of accounting." There was also a situation where a most prestigious car and jet engine maker found itself in great financial difficulty with a marked deficit in its operation. After hiring an "expert" who looked at the books, the company instantly became, without any effort, profitable after all. Accordingly, all it needed was to capitalize on some of the obvious R&D expenditures. No wonder someone once said: "True, accountants can show us all kinds of profit, but we don't believe them."

Profit as Economists See It

In as much as the discipline of economics involves resources allocation and utilization, the issue of profit has been and always will be the focal point of economic analysis. Profit applies to individual economic undertakings, whereas for a national account, the term economic growth is being popularly utilized. However, to define profit for the economist is not a simple matter. In fact, ever since 1776, when Adam Smith's *The Wealth of Nations* was published, no economist could show the public just what profit is.

Adam Smith's economic treatise established the ideal of free choice which has become a foundation of moral philosophy. Hence, Adam Smith is not only to be respected as the founding father of classic economic analysis, but also a major philosopher. As a philosopher, he had not in any way made clear just what profit is, even though, from the beginning of his deliberation he made himself clear enough that wealth consists of money, and every commodity should be expressed in monetary terms. He continued that:

> A frugal man, or a man eager to be rich, is said to love money; and a careless and generous, or a profuse man is said to be indifferent about it. To grow rich is to get money; and wealth and money, in short, are in common language, considered as every respect synonymous.

What a tragedy that this great man led us to believe! In the first place, there is a world of difference between being rich and wealthy. A person may use money to buy his or her freedom, but it is unlikely that he or she will sell his or her freedom for money, since without freedom money is useless. Adam Smith had discussed at great length money and wealth, but had established no clear notion about what is profit. The closest he ever came to say is that:

> The whole drugs which the best employed apothecary in a large market-town will sell in a year may not perhaps cost him above thirty or forty pounds. Though he should sell them, therefore, for three or four hundred, or a thousand per cent profit this may frequently be no more than the reasonable wages of his labour price of the drugs. The great part of the apparent profit is real wages disguised in the garb of profit. In a small seaport town a little grocer will make forty or fifty per cent, upon a stock of a single hundred pounds, while a considerable wholesale merchant in the same place will scarcely make eight or ten per cent, upon a stock of ten thousand (Marshall, 1920, p. 506).

Alfred Marshall, who advocated his notion of profit much the same as Smith, wrote:

And, if a man employs in business a capital stock of goods of various kinds which are estimated as worth of 10,000 pounds in all; then 400 pounds a year may be said to represent interest at the rate of four per cent, on that capital, on the supposition that the aggregate money value of the things which constitute it has remained unchanged. He would not, however, be willing to continue the business unless he expected his total net gains from it to exceed interest on his capital at the current rate. These gains are called profits (Marshall, 1920, p. 62).

In the process of profit analysis, economists have long recognized that there is a great deal of confusion between someone who runs his or her own business, whose entitlement for his or her labour is inseparable from the profit the company earns, and the case in large companies where managers are paid employees, and profit is then a residual of revenue after deducting all expenses. There are also other difficulties. The noted ones that seem to have no apparent solution are:

1. The production period, as production is a roundabout process, once begun, it will be just like drawing a circle, with no beginning or end. Of course, if the business operation terminates then the production period as well as profit can be determined.

2. The issue of the aggregation of all economic activities through the common denominator of money. It is obvious that not all economic activities can be assigned value in monetary terms. For example, in the case of risking your health or family happiness for business. Can such a risky undertaking be reduced to dollars and cents?

While there is difficulty in finding a universal notion about what profit is, the most popular one in use is perhaps J. R. Hicks' idea of profit. Profit, according to him, is when an individual is as well off in the end as he was at the beginning (see Hicks, 1992, p. 172). His definition is simplified by the author for illustration. Accordingly, a profit can only be possible if an economic undertaking is as well-off in the end as the undertaking was at the beginning. Hicks also said: "It cannot be profitable (in this sense) to make machines unless the use of the machines is also profitable; so, to assess the profitability of investment, we should look right forward to the production of the final product" (see Hicks, 1992, p. 139). It is obvious that in Hicks' mind making a profit is a value-added process, but the value-adding process should not be merely for the individual, but for society as well. Incidentally, the person who said: "Accountants can show us all kinds of profit, but we don't believe them," also said: "Whilst economists can tell us all about profit, they cannot show what profit is." Who knows, perhaps there is some truth about his observations on both.

If we are to assess what has been said at the beginning of the chapter, and attempt to further our illustration about "profit", we may derive the following:

1. Our resources are limited in supply, and to use such resources for any purpose will yield no profit. Resource-rich countries through the exploitation of resources for consumer goods and services yield no profit, unless, the utilization of such resources can generate greater value. Only then, can there be a profit.

2. As industrial growth takes place, countries are so proudly announcing their high rate of economic growth without taking into consideration the replenishment of resources used to make that growth possible. There is no economic growth for these countries. At best, they have impaired their capital assets supposedly needed for future production needs.

3. Taking the world as a whole, it matters not whether countries report economic growth or recession. Economic growth means "profit" (as well-off at the end as at the beginning), which in fact is a measurement of how we utilize our natural resources. If there is high level use of the limited resources, it is a matter of how we use our scarce resources. If we use resources responsibly, there will be more for the future. On the other hand, if we use resources ruthlessly or irresponsibly in the name of making "profit", then, there is really no "profit" at all, just a ripping off from future generations for the present few. Where is the profit if a country that uses nuclear power to generate electricity has no way to dispose of nuclear waste?

Accountants and economists are fully aware of the scarcity of natural resources, yet both fail to make any attempt to account for the utilization of these resources (both the utilization and damaging effect). Therefore, profit derived on the basis of theorized economic analysis and accounting practice is not profit at all, because neither consider the cost to the environment before profit is derived.

It matters not how profit is defined by words or manipulated by figures. What matters most is for economists and accountants to communicate information responsibly. The rest is up to the managers of the company on how they conduct their business in the interest of those who trust them, including our future generations.

If words are not being played with, is there such a thing as "profit"? The answer is no since we cannot restore limited resources. But we could have operational residuals, and such residuals must be derived by accounting all inputs utilized including the depletion of environmental factors, so business managers will be able to manage companies more sensibly and responsibly. On the other hand, if a question is raised in respect to the goals of firms, and if what they are here for is not profit, what would their goals be? The answer would be to allocate resources responsibly for the purpose of creating wealth for the individual and adding value to society.

3

A survey of
management concepts
and theories

The theory of the firm emphasizes profit and departs from the fundamental function of economic analysis that requires all economic activities to be directed at allocating resources for the purpose of creating wealth and adding value for the individual and to society. And it is the goal of making profit for the firm that prevents the new thinking process needed for business management to break the knowledge barriers.

Corporate Managers: Chasing Rabbits in the Forest or Killing Demons in Computer Games

A corporate manager is the star of his or her domain. Ideally, he or she should be in full command of the resources allocated by the corporation (the entity) for the purpose of fulfilling the firm's objective. But tasks confronted by corporate managers are far from simple. If the firm's objective is profit, management efforts will be devoted to the attainment of the objective. Anything that hinders profit-making will become a problem to management. Consequently, the management challenge will likely be narrowed down to two courses of action: attempting to make profit, which is very much like a hungry wolf chasing rabbits in the forest; solving problems, which resembles killing demons in a computer game – the more demons killed, the greater the onslaught. In addition, there are always people who claim that chasing rabbits and killing demons are not the only challenges, that there are other invisible "lords from the hedge" – environmental protection, defenders of employees' rights and last, but not least, ethics! What a nightmare! On the other hand, no one ever said that being a corporate manager was an easy task! Perhaps that is why so many people have advice for corporate managers on how to manage a corporation successfully while bettering themselves (whether to get a free parking space in front of the office building or a huge salary, expense account or stock option; they are all part of the package) and at the same time making the world a better place to live in.

In the old days, learning how to manage a business was a process of apprenticeship. The institutions of higher learning did not recognize it as an academic pursuit but considered it to be a trade education. However, business management courses were included in the commerce programme, with most teaching in business based on tested theories. But it was after the Second World War that the fashionable business school found its own place in the university environment. And it was also about the same time that there was a shift in teaching focus, away from management theories to case materials. However, regardless of how case studies have infiltrated the management educational system, theories have always made their way into the learning process.

Management Theories and the Monkeys

A theory is a logical argument intended to explain facts or events. In these days, a good theory should be tested and applicable to real life situations. For example, in a recent trend, a theory advanced by D. Charles Handy, a British organizational theorist, advocated that "enterprising workers could leave their jobs and become portfolio workers instead, selling their skills and services, like freelance workers, to the highest bidders" (Ho, 1995, p. 5). There is no statistical evidence to prove his theory is workable, yet in the US and Canada, freelance workers seeking work from various sources is common practice, and not exceptional (consultants, for example).

Another theory is that humans evolved from the same ancestors as monkeys. This theory cannot be proven (since we cannot go back and replay history, paleontologists have worked the assumption for quite some time. It was not until 1921, a half century after Darwin first published his work, that the discovery of the Peking Man (a skeleton some 250,000 to 500,000 years old which appears to be an extinct precursor to man) was made known to the world, and the theory was placed on firmer ground (though it is still not accepted by some). Most of us have learned this theory, though there are those who hold other beliefs, such as one where our first parents committed a mortal sin by eating the forbidden fruit that gave us wisdom, but drove us out of paradise.

So monkeys are our cousins. What this theory does is establish some basis for research into fundamental behaviours that we might share with monkeys, so that from the study of monkey behaviour, we learn more about ourselves, since we cannot put humans in the research laboratory. Illustrating this principle is the following experiment, for which unfortunately, the author does not have the sources, but which was shown on television some years ago.

> A group of monkeys were gathered under a tree, to which the researcher had strung a banana high enough the monkeys were unable to reach it. The experiment was intended to test the monkeys' intelligence and assumed that the most intelligent monkey would find a way to get the banana. Sure enough, as soon as the monkeys spotted the banana, a big commotion started; some jumped around, others raced impatiently around the compound, and a few did nothing but walked around impatiently. But one monkey, watched the banana quietly for a while, then climbed the tree, got to the branch to which the string was tied and pulled it up to get the banana.

The researcher concluded that this particular monkey used his brain and got the banana. That was the theory, and the theory had been tested. All theories are used to guide human action. Management theories have done their bit to shape our economy through all commercial undertakings. Corporations through their managers have done well in shaping the commercial world and services to people at large (consumer driven, so to speak), but at the same time, there have also been serious concerns about corporate conduct, and among other things, the fundamental purposes of the corporations were questioned as well.

What is Wrong with Corporate Management Practice?

It has taken a long time, but most human beings have finally come to the conclusion that there is nothing to be gained if we continue to wage wars in order to get what we want. Instead, we have a new sense of belonging ("we are all part of planet earth," as said by Neil Armstrong during his space orbiting: "Oh! Earth, our home!") and are learning to work together for economic reasons. Although there are many facets to a market economy, the success of it all, if we can put politics on hold, very much depends on how harmoniously we can work with each other and our environment. This applies particularly to the major players in the market economy, who are none other than the large corporations which must be efficiently, effectively, meaningfully and responsibly managed.

For the last century and especially, since the end of the Second World War, there have been enormous advances in science and technology. The discovery of the computer chip and the extensive use of electronic technology, and more recently the internet explosion have created an alternative economics of innovation which has not only changed the face of business communication and information processing, but also brought along a change in the mindset of people, created social change, and altered the industrial structure of our business world. For example, a home-based business can use free software to access internet, monitor, acquire, and use the technology developed around the world with an efficiency that large corporations need to learn from. Looking at the global environment, there is the formation of the European Union (EU), North American Free Trade Agreement (NAFTA), Association of Southeast Asian Nations (ASEAN), Asia-Pacific Economic Cooperation (APEC) and the World Trade Organization (WTO), all designed to lower and eventually eliminate trade barriers, so there will be free flow of goods, services and maybe even labour. That will be the day when human beings finally understand the statement: "Oh! Earth, our home!" These changes must give our corporate managers a new sense of urgency, and purpose in their managerial life.

Taking business management as a whole, unfortunately, whether we like it or not, nothing much has changed for at least fifty years. Even the newest idea of "intrepreneurship" is nothing more than an extension of the "profit centre" concept that has been on the upswing since the 1940s and now has a new coat of paint. We can temporarily forget about Theories X and Y, and even the new kid on the block, Theory Z, an American based theory modelled after the Japanese way of managing, but without the Japanese culture. It is all very much looking like the book on the professor's bookshelf with its pages turning yellow, which will need more than just a good dusting to make it look new again. Then, an honourable mention must be given to our two grand old ladies – system thinking and customer driven, for their reappearing act under the banner of "Quality Management".

What is wrong with our corporate management? Plenty.

1. Profit maximization and/or ROI as the goal of the firm has resulted in an approach to corporate management which has hampered managers' ability to think positively and erode individual's initiative and willingness to create and innovate.

2. The use of dehumanized management strategies that squeeze humanity out of the organization.

3. Looking for solutions through "intrepreneurship", but with failure to recognize that if the corporation is an entity of its own, and shareholders are owners of the entity, then without ownership, everyone working in the entity would be an employee, including the CEO, and therefore, the entity is entrepreneurless. How can you have entrepreneurship if the firm is entrepreneurless?

4. Last, but not least, there is the flaw embedded in business management concepts which is caused by the departure from the fundamental function of economics by the theory of the firms. The theory, instead of focusing on directing resources to create wealth and value for the individual and society, concentrates on making profit for the firm. Consequently, such development caused conflict between (a) accountants and economists on issues of "profit" (for more details, see Kao, 1995, pp. 34–50); and more importantly (b) the interests of the firm and the common good. Moreover, and of crucial importance, it prevents the new thinking process required to break free from the knowledge barriers

which have restricted management theory for so long. This flaw drives the corporate managers knowingly or unknowingly in an endless search for short-term profit (in the extreme, for instant profit) at the cost of the "common good": the people, resources and the environment.

The Corporate Management Challenge

The bottomline approach to corporate management

The bottomline approach to corporate management is no secret to anyone involved in corporate life. It could mean profit maximization or a ROI to please stockholders and induce investment (capital driven). However, both profit maximization and ROI have had their days. Not only do people working in the corporate structure have doubts in their minds, but those who are knowledgeable in economics have also raised a lot of serious questions.

The theoretical development of the idea of profit maximization comes perhaps from the microeconomists, since the theory of the firm has become the bible of microeconomics learning and teaching. It has been assumed that a firm is an economic entity, and profit maximization is the goal. Ownership belongs to the investors (stockholders), and anyone who works in the entity is an employee, working for the achievement of the entity's goal which is profit maximization. To many of us, this ideal sounds good, since few have ever questioned the nature of profit, nor how it should be distributed in a capitalist society. This idea of profit maximization for the stockholders' satisfaction was further compounded by another idea invented by some "B" (Business) school academics, implying that the best way to make money is to use other people's money. Therefore, profit maximization seems to be the goal of firms, the goal of the managers and everyone working in the entity. This conventional wholesale wisdom might make some sense to investors or patrons of microeconomists, but there is a great deal of doubt about its validity.

A few observations by Nobel laureate Herbert Simon are a case in point (Simon, 1967, p. 9): Just as the central assumption in the theory of consumption is that the consumer strives to maximise his utility, so the crucial assumption in the theory of the firm is that the entrepreneur [note: it refers to an entrepreneur, not a corporation] strives to maximise his residual share-profit. Attacks on this hypothesis have been frequent. The important ones may be classified as follows:

1. The theory is ambiguous – is it short-run or long-run profit that is to be maximized?

2. The entrepreneur may obtain all kinds of "psychic income" from the firm. But if we allow "psychic income", the criterion of profit maximization loses all of its definiteness.

3. The entrepreneur may not care to maximize, but may simply want to earn a return that is regarded as satisfactory.

4. It is often observed that under modern conditions the equity owners and the active managers of an enterprise are separate and distinct groups of people, so that the latter may not be motivated to maximize profit.

5. Where there is imperfect competition among firms, maximizing is an ambiguous goal, for what action is optimal for one firm depends on the actions of the other firm.

The ambiguous nature of profit maximization imposes no real meaning to business itself nor to the people who manage it. To corporate managers, it is the challenge of how to manage the business that creates wealth for the individuals involved in the business and adds value to society.

Along the same vein as the profit maximization approach to corporate management, there is the matter of the use of ROI as the goal and management objective of the firm. Like profit maximization, ROI is just as ambiguous, if not more so, since it is simply an arithmetical calculation, subject to accounting manipulation. It ties in with capital investment, and investors are mostly interested in making money (high earnings and high turnover) rather than in the business itself. Therefore, the purpose of ROI being the goal of the firm, is not only to please existing stockholders, but is also intended to attract potential investors to invest in the firm. If a firm's shares are traded in the market, ROI affects the trading price. It is also for this reason that the methods used to determine a firm's profit, and hence, ROI, are closely monitored by the stock exchange, not on a long-term basis, but based on current activities. Consequently, as much as the accounting practice works to ensure that the stock exchange commission will not raise any doubts about accounting practice, corporate managers also monitor their firms' stock price traded in the exchange as an indicator of their performance. This is done not only on a year-to-year basis, but at least quarterly, and more often than not, weekly or even on a daily basis. This short-term approach for goal attainment can only motivate managers to plan, make decisions and act for short-run gain and cosmetic improvements. It forces them to steal from the future for present day performance rather than encourage them to think and plan for long-term benefit for the firm, society and humanity. The ROI approach tends to motivate managers to behave like short-sighted dinosaurs, eating up whatever is in front of them without thinking of the future and driving themselves to extinction.

As cost determination is directly related to the size of profit, without taking into consideration environmental costs, the cost of doing business is seriously understated. So far, neither economists nor accountants have done much to account for these costs. Therefore, the so-called "profit" earned by our corporations is being seriously overstated. If the overstated profit is distributed to a few rather than recycled for the preservation of environmental health, it could mean that our corporations are taking what belongs to the future to make a few "rich" now. This is not a matter of justice, but an unfortunate reality that must be attended to. Otherwise, it is likely that we will be taking our children to the grave with us, since we will have used up their life support for our own excessive consumption.

Despite the fact that economists and accountants make no attempt to communicate environmental cost information to guide managerial actions, corporate managers must become aware of this challenge, and find ways to get out of this short-run, more seriously, "instant" mentality. There are many future needs to attend to, not just to satisfy the investors' urge to make money, money and more money for their personal consumption and enjoyment, without regard for the consequences on the future. However, some managers would say: "Can you reach the promised land if you are dead?" This, of course, becomes the challenge to corporate managers: can their performance satisfy the stockholders' desire for ROI attainment, while benefiting the firm, society and humanity in the long-term?

The matter concerning dehumanized management strategy

The cause of the dehumanization of the corporate "entity" is nothing other than concept itself. Under the entity concept, everyone in the entity is a part of the working machine. Work is to be standardized, rules and regulations must be observed, procedures must be followed, and staff control must be rigid. Should there be a grievance, rather than sitting down to deal with the matter like normal human beings, the grievance procedure has to be followed: the matter has to be taken up by the shop steward who deals with the supervisor who takes it to the manager and so on in a step by step process. The following is one incident which the author had the personal "privilege" of experiencing.

A manager passing through a stockroom of a large corporation's warehouse picked up some trash from the floor and tossed it into a bin. This was noticed by the shop steward who filed a formal complaint against the management because cleaning up is a classified, paid union job, and should be done by a union member. What the manager did was a serious breach of the union-management agreement. The circumstances prompted a crucial question: Can there be any management, if there is no evidence of trust?

In a dehumanized institution, if a decision is required to be made, the right person has to be found to make the decision. So even if a fire is raging, if putting out the fire is not in the manual, then must we allow the fire to burn?

In reality, things may not be as rigid as described above, but if a decision is made outside of the prescribed manual, or the area of authority, who then has the nerve to make the needed decision, if there is a fear of punishment? This is known as "management by fear". All in all, in the dehumanized institution, human beings are no longer decision-makers. The only decisions are programmed decisions; these could just as easily be made by a computer, and as an added incentive, a computer can and has less chance of making mistakes. On the other hand, computers can be wrong as well, but who can say that the computer is wrong if the only person who knows how the computer made the decision is the programmer?

Here is a little story about what might happen in a dehumanized organization and dehumanized society where everything is decided by computers.

In the future, computers will programme everything humans have to do, including job allocation. One person who works for a construction company is assigned by the computer to work one day a week, to push a button for some important project. This of course takes almost no time at all, but if he were to do more then it would mean taking jobs away from others. One day, he rose early in the morning ready to go to work (to push the button) and happened to pass by the museum where he saw a spade and a few other construction tools on display in a glass case. Suddenly he was seized with the desire to taste what it was like to work with a spade on the construction site. He broke the glass case and took the spade. Just when he was about to get into some old-fashioned manual labour, he was arrested, not by a human, but by a computer. The computer took him to the municipal court, presided over by a judge who was also a computer, where he was tried for aberrant behaviour and destruction of public property. He was found guilty and sentenced. The sentence: his right to work the push button once a week was taken away for a period of three months.

The above may never really happen; after all we have not yet in our corporate structure dehumanized the organization to this extent, but it remains a valid illustration of our current management trends. Is this what we want business management to become?

The challenge to the corporate manager is obvious – remember that we will always be humans with human dignity. The dignity to make decisions for ourselves, for the corporation, for society and for humanity as well. One of the latest trends is the use of computer-run telephone message systems. Between waiting on hold and being transferred from one synthetic message to another, it is easy to spend a good three quarters of an hour (or more) on the phone, with no human contact. What are humans for anyway?

Perhaps just to be used for a voice recording to prevent human contact (or intrusions from space)? Technology advancement is supposed to facilitate our lives; instead, it is used to facilitate the business entity, and we are forced to adapt to it. Management is unable to foresee the consequences of technology; it has not improved efficiency, nor effectiveness. If the goal of a corporation is to become a "perfect entity", attainment of this goal may see it become not only entrepreneurless, but humanless as well. Just do not forget that we are and always will be humans with human dignity; dignity to make decisions for ourselves, for the corporation, society and humanity as well.

And another thing, everywhere you go, slogans remind people in the organization of the corporate goals, and that everyone must work as a team to achieve the goals of the corporation. But not a word is said about the individual's goal. Don't humans count any more? Or have they never?

The intrepreneurship story

"Intrepreneurship" is a new word, and you may not find it in a dictionary. From the corporate management's point of view, it is a change, if nothing else, a change to announce a new word to the disenchanted concerned people.

Digging into its roots, the idea behind intrepreneurship is to give a division (or unit manager) some degree of decision-making authority, in order to make some tough decisions (particularly long-run management decisions) and live with them. If intrepreneurship was examined closely, it would not be difficult to find that underneath the cosmetics and a thin new skin, the blood is the same as that of its biological father, the "profit centre" concept. It has been in the market for quite some time, and the profit centre concept is subject to the same old ROI jingle, with decisions being made to attain corporate goals and please the stockholders, and thereby the stock exchange. So why mix it up with intrepreneurship when entrepreneurs are supposed to make decisions for themselves?

Intrepreneurship has another connotation largely referring to inventors in large corporations. The idea is to allow the inventor to form a small company (with his or her invention) as a spin-off from the parent structure. The trouble is, will the parent company stay on in the minority position, so the inventor can run "his" or "her" company without consolidation of its operational result? The answer is often no deal. In this case, the situation is no different from running a subsidiary for the parent company, and subject to corporate staff control. It has to be clear that if staff control is applied to subsidiaries, there will be no entrepreneurship. In all cases, a staff function is supposed to provide expert advice (or service) to assist the operational personnel to achieve the task. More often than not, the staff function becomes a control mechanism, a watchdog for the "boys" or "girls" on the 47th floor who have the "executive washroom keys". It is a bureaucratic process of pushing papers and following procedural guidelines and above all, making sure that everyone follows the "dotted lines".

A Survey of Management Theories

Corporate (or business) management is a big topic and it has a big history. The way our corporate managers manage business today is not a miracle that happened overnight or across the boardroom table. It evolved with time and societal, political, cultural and professional pressures, and, lest we forget, the looming presence of stock exchange influences. Before getting serious about an entrepreneurial approach to corporate management, it would be useful to look back at some theories (and thoughts) developed over time and how they relate to today's business and human environment. This will lead into the theory of entrepreneurism, advocated by the author on the basis of the creative nature of humankind. In this theory, management function is regarded as working with people to manage limited resources responsibly for the purpose of creating wealth for the individual and adding value to society.

Before the Industrial Revolution, people in Europe were bound to their station in life, and business management was primarily dominated by cultural values. Matters were similar in China, where merchants and tradesmen were forbidden to wear silk and the community was anti-business, anti-achievement, particularly in commerce and trade. Kings and queens were the rule, and they ruled with the help of faith, and the actions and discretion of powerful individuals. In fact, there was little formalized thought on business management with the exception of some Confucian ideas which originated about 2,000 years ago. It was not until the arrival of the industrial era that a wave of philosophical writings characterized by discussion of ethics or standards for governing human market behaviour finally emerged and shaped a new society, where individual gains and achievements were recognized. These and other things made it necessary to search for a more rational and systematic knowledge base for business management. This evolution marked an advancement of business management theory, although not to the same scale as seen today. The change nevertheless led to a new business cultural environment.

In England, the need for factory workers for the Industrial Revolution led to the breakdown of families, as children and women were forced into the labour market. At the same time, the demands of new large-scale production processes prompted the need for re-organization to meet production requirements. Consequently, there was a need for the infusion of capital, planned organization and control measures to achieve the predicted performance. Systematized management became a necessity in order to cope with business expansion and competitive challenges.

Most of us are told that Frederick Taylor (1856–1915) pioneered "scientific management", but before his time, Charles Babbage (1792–1871) had already introduced scientific management thought, though this has not been formally recognized. He demonstrated the world's first practical mechanical calculator, and in 1832, he introduced the "difference engine", the direct ancestor of the "Burroughs" accounting machines, developed some ninety-two years later. His most successful book was *On the Economy of Machinery and Manufactures* in 1832. Like Adam Smith, he was deeply interested in the principles of the division of labour. For Babbage, the division of labour brought more efficiency because of reduced job training, reduced material waste, elimination of time waste from occupation changes, and among other things, improved skills development and the use of tools. Although management thought could have been formalized at that time, management was more concerned with pragmatic issues such as finance, production process, selling, and obtaining labour, all of which were at a critical stage, rather than developing principles or generalizations about management. It should be noted that it was Robert Owen (with Richard Arkwright) who led the way in pre-planning factory layout and who wholeheartedly searched for a new harmony between the human factor and the age of machines (Morrison and Morrison, 1961, p. 9). Following soon after was the early development of management thought along with industrialization, mass production and the factory system. It was about the middle of the nineteenth century that management theories emerged, becoming a more formalized body of knowledge. In general, there are four major groups of management theories (see Table 3.1) (Tann, 1970). Classical management theory embraces scientific management and bureaucratic management. As advocated by Frederick Taylor (1856–1915), scientific management aims to standardize work methods and use rational selection of employees coupled with training and job development to achieve efficiency. Bureaucratic management was advocated by Max Weber (1864–1926) as the way to deal with management challenges for large corporations such as General Foods, IBM and Standard Oil. The model organizes massive numbers of people in a complex situation. The bureaucratic model relies on multiple levels of reporting to divide responsibilities and control resources. The unfortunate result is often a pyramid of committees and subcommittees, endless paperwork, meetings and more meetings.

The more sensible group of management theories stresses the importance of human relations. It is a good thing somebody thought about the people working in business. This movement has been the result and efforts of a large number of individuals and has made a remarkable impact on management theory and practice. The bottomline is still super performance in the attainment of the goal of the firm, but human beings now receive a significant share of management's attention.

System theory came along with the advancement of technology. Theorists in this stream advocated that a system is a collective association of interrelated and interdependent parts. Machines and technology are part of our system, but systems are defined by relationships among people. The systems approach provides a frame of reference for managers who must make decisions in a constantly changing environment.

Table 3.1 Summary of management theories

Name of theory	The theory	Who's who
Classical	Scientific management	Charles Babbage Frederick Taylor Lillian and Frank Gilbreth
	Bureaucratic management	Max Weber
	Administrative management	Henri Fayol
Behavioural group	Human relations: better relations, better performance	Chester L. Bernard Kurt Lewin Mary Paeker Follett Elton Mayo
	Theories X and Y	Douglas McGregor
	Motivation theories	Abraham Maslow
	Integration theories	Victor Vroom Lyman Porter
	Theory Z: American adaptation of Japanese organizational behaviour	William Ouchi
System	Contingency theories	Frederick Lurhans
Quantitative	Operations management Management science Management information systems	The use of computer technology, model building and mathematical manipulation

Source: Modified from Kao (1995, Table 4.1, p. 56).

Contingency management suggests that managers should be prepared to adapt their leadership role to accommodate different situations. On the other hand, it is also part of the theory that managers should be assigned to situations that best fit their leadership style.

There are also other management theories, including the recently developed quantitative management theory which is represented by management science, operations management and management information systems. There is also the Type Z theory group, an American adaptation of Japanese management practice, which focuses on the improvement of management towards employees' interests with measures such as long-term employment, improved benefits, and most important of all, concerns for employees and their families. The idea of concern for employees and their families is a good one, except there is still a fundamental difference between Japan and the US. In Japan, management efforts are devoted to developing individuals as members of the corporate family, but in the US (and possibly other countries), management effort tends to develop individuals as employees.

A short introduction will also be given to the Theory of Entrepreneurism, developed on the basis of the author's earlier work, *Entrepreneurship: A Wealth-Creation and Value-Adding Process* (1995), and applied to corporate management. (See Table 3.1.)

From Theory of the Firm and the Theories of Management to the Theory of Entrepreneurism

The theory of the firm was adopted by people who needed a conceptual base to challenge Karl Marx's *Das Kapital*. Capitalism is labelled by communists as a vampire which sucks the blood out of the mass of the working population. Now that the Berlin Wall has been torn down and communist and socialist regimes have fallen all around the globe except in China, Cuba and one or two other countries, capitalism has experienced a worldwide resurgence. All of a sudden it has become a universal panacea which will bring wealth to all. In this capitalist utopia, people will enjoy the same standard of living everywhere, and the price and availability of goods and services will be the same in France, Thailand, China, Costa Rica, Reunion Island, or the Ivory Coast. Will the practice of capitalism really do all this for us? This is the trillion dollar (here again, the reference to money) question. Nevertheless, no matter how we see it, equality or no equality, capitalism or no capitalism, the name of the capitalist's game is still corporate management.

We have a line up of management theories used to build models and to guide actions which have to be applied and implemented by people. All theories and models are developed on assumptions, and the management model and theories of management assume the firm is an independent entity. It is an assumption derived from the theory of the firm, and if the origin is to be traced, the credit should be given to Adam Smith, even though Smith had nothing to do with the theory of the firm. Indeed, if he could be informed about what the theory of the firm has done to his "Wealth of Nations", it would make him cry in his grave; his theory was about allocating and directing resources to create "wealth", not to maximize profit for the firm.

To appreciate the development sequence, we begin with Adam Smith's open market concept, continuing through to the theory of the firm. We then move to general management (more details in Chapter 4) and how the accountants made ROI the goal of the firm. This will lead to the theory of entrepreneurism, an ideology advocated by the author designed to recognize the creative nature of humans, and make management work with people to allocate and direct resources for the purpose of creating wealth for the individual and adding value to society (see Table 3.2).

A Short Glance into Japanese Culture and Japanese Management

It is the success of Japan's post-war economic performance that has made the whole world curious about the secret of Japan's success. Learning how the Japanese manage their businesses (including morning exercise routines for factory or store employees) has been, for quite some time, a fashionable thing in the Western world. The development of Theory Z is a case in point.

To know about Japan's success is not an easy task, because we need to know how the Japanese feel about life, the world, and above all, their mindset. To understand the Japanese mindscape, it is useful to see how it developed through some 10,000 years of the Jomon, Yayoi and Yamato cultures (Maruyama, 1982).

Europeans often trace their historical development to ancient Greece of some 2,400 years ago; similarly, the Japanese mindscape can be traced back to their historical development of some

Table 3.2 Sequential development of theories in economics, management, the management model, ROI and the theory of entrepreneurism

Development sequence	Theories and practices	Utility
1	Adam Smith's open market economy: the market is governed by supply and demand.	The founding of classical economics.
2	The theory of the firm: the firm is an entity, the firm's goal is profit maximization, individuals in the firm have the same goal as the firm. The theory separates people from the firm making the firm entrepreneurless.	Made it possible for the development of business management into a profession. It induced the development of management theories.
3	Theories of management: the goal of the firm is profit maximization. Management is a function in the market economy, and management resources need people to achieve the goal of the firm.	The basis to build models; guide for actions.
4	The general management model: advocates the notion that a firm is an entity, management to be separated from ownership, a profession specializing in managing other people's business.	The basis of curriculum development for business management education and guide for action.
5	Strategic management: management in action, mapping out goal attainment strategies within available resources and the existing environment.	Activation of theories and models through action and managing.
6	Rate of return on investment: an arithmetical calculation that bridges profit, the theory of management, general model and strategic management. Its universality stands worldwide.	A simple tool to measure a firm's performance which made capitalism a living giant that dominates the capitalist economy.

Table 3.2 (cont'd)

Development sequence	Theories and practices	Utility
7	The theory of entrepreneurism:* a theory advocated by the author. Focusing the management's function; all human endeavour to create an entrepreneurial society which is one in which the economic activities are directed at creating wealth and adding value for the individual and society.	Based on the human need for ownership, and transform it to model to stimulate the individual's need for creativity and innovativeness as the basis for corporate management.

* To the author's knowledge, the term "entrepreneurism" was first noted from the title of a book by Thomas F. Jones (with T. P. Elsaesser), published by Donald I. Fince, Inc., New York, 1987. It is a book about how to start a new business. The term was subsequently mentioned at a Babson College Conference in the early 1990s, with the same connotation.

Source: Modified from Kao (1995, p. 58).

2,400 years ago. The Japanese mindscape can also be traced back to the three main sources of the Japanese culture: the oldest, the Jomon culture that began 9,000 years ago, the Yayoi culture, the homeostatic one that began 2,300 years ago, and the hierarchical Yamato culture that arrived via Korea about 1,500 years ago.

> The Japanese people express their philosophy more readily in various art forms and design principles than in verbal discourse. Much philosophical thinking particularly went into designs of gardens, architecture, and flower arrangement. It is from archaeological and anthropological studies of such designs that we gain an understanding of the development of the Japanese thought patterns (Maruyama, 1982).

Some of the basic principles of Japanese garden design and flower arrangement are: (1) avoidance of repetition; (2) harmony of dissimilar elements; (3) interrelationship between heterogeneous elements. While outsiders do not always appreciate these principles, even Japanese gardeners and flower arrangers are not aware of these principles, because they are taken for granted. It is not at all surprising to know that there are more than 2,000 different "schools" of flower arrangement in Japan.

Japan is limited by its space, the consciousness of space allocation has been and always will be in the minds of the Japanese people. Therefore, efficiency is not some management way of shouting slogans and putting up posters to make workers aware of the importance of being efficient; in Japan it is in everyone's system. The cultivation of rice, and the concept of land use, both have their part in Japanese thinking and have impacted management thought and practice. Japanese management style is rooted in Japanese culture; it is for this reason Westerners, and the rest of the world, should not copy Japanese management. Comparing Japanese and American management, Maruyama suggests the following (1982, p. 68):

1. The Japanese way of management has an epistemological basis different from what is called "participatory management" in the US.

2. The future principles of US management should be neither an imitation of the Japanese way, nor the traditional American way. Some of the important principles are: (a) recognition of heterogeneity as an asset, not liability; (b) combination of heterogenous individuals into mutually beneficial interaction patterns, which Maruyama called "morphogenetic management".

For most of us who wish to know more about Japanese business management, it is necessary to understand their language, as well as their historical background. A typical Japanese businessman (Japanese business is very much dominated by men) always has the interest of his country and countrymen at heart. For instance, CEOs of Japanese corporations will drive a Toyota Crown, the most luxurious and prestigious car of the Japanese range. They will never choose a German car such as Mercedes Benz, or a Jaguar Sovereign made in the UK. This tradition is perhaps inherent to what the Japanese learnt from the Confucius doctrine which emphasizes that business must be guided by "Zen" (perhaps it was Zee-Kon), which is to be passionate and good to your fellow businessmen but unfortunately, only half of this wisdom is practised by them.

Although the Japanese businessman is very passionate towards another Japanese businessman, they are completely ruthless to outsiders. On the other hand, when the Canadian Government signed the free trade agreement with the US, there was a shift in market trends, and Canada desperately needed businesses to provide jobs. Instead of helping their countrymen, a large number of Canadian firms moved down South to cash in on what we called lower "production cost" or a "bigger market". This included the then Chairman of the National Entrepreneurship Development Institute, who moved his plant to the US. When the author asked him why, he said: "Raymond, it is a business decision."

Bear in mind, it is a known fact that the Japanese economy has not been doing well all the time; every day is not the 4th of July. At the tail-end of the twentieth century, Japan has seen prolonged recession caused by capitalistic expansion, which has resulted not only in loss of jobs, but also a high small- and medium-enterprise fatality at home, and a near bank credit crisis. These are all visible signs that it matters not whether it is Japanese management or some other form of corporate management; there is a limitation on capitalistic expansion. Japan's economy could only survive or prosper through its corporate expansion on the basis of continued access to the cheaper resources – both material and human – of other countries. The same applies to other industrialized nations, including matured nations such as Germany and the US. The management challenge will not only be in marketing – to sell goods and services – but more so in the handling of human and natural resources. One indication of the future: One of Japan's giant car manufacturers is in fact no longer manufacturing cars in Japan, but expanding its car-making business throughout all parts of the world.

It should also be noted that in the Japanese corporations, individuals in the organization are no more than computerized functioning units. Absolute obedience is required among individuals working in the factory, deviations or changes made in any process is forbidden, and serious situations are subject to reprimand or dismissal. If "creativeness and innovation" are the essence of entrepreneurship, then it would be extremely difficult to find any "entrepreneurial culture" in the Japanese working environment.

This is perhaps why some Japanese top executives are concerned about their working team no longer being "entrepreneurial". How can there be any entrepreneurs if decision-making options are not part of the human assets?

Quality Management

In coping with technological advancement, people involved in the academic world are also pressured for performance. Unfortunately, we have become used to the capital driven management mode for the attainment of "ROI" and "short-term profit". And everything we do we want to get "instant results" or "instant success". As a result, some academic researchers and writers, instead of breaking new ground, have taken a short cut to paradise, by coming up with the concept "Total Quality Management" or "TQM" as we know it today – and believe it would make the difference.

The definition

Although TQM has been in fashion for the past decade, in comparison with some old concepts such as MBO (management by objective), MBE (management by exception), MBC (management by crisis) and MBF (management by fear), TQM is still the new kid on the block. For better or worse, the new kid has already caught the attention of academic curriculum planners. As such, a number of "B" schools have started to offer a course or two in their MBA or undergraduate programmes or both. What puzzles the author is that to this day, he finds it difficult to see the "quality" element in the "Quality Management" or the "newness" in this new kid.

TQM has different versions in accordance with how an individual perceives and interprets "Quality Management". A more recent one and cited by a whole lot of people is the definition of TQM provided by Coat (1990):

> Total Quality Management (TQM) is a commitment to excellence achieved by teamwork, and a process of continuous improvement. TQM means dedication to being the best, to

delivering high quality services which meet or exceed the expectations of customers.

Figure 3.1

(A) Problem-solving:

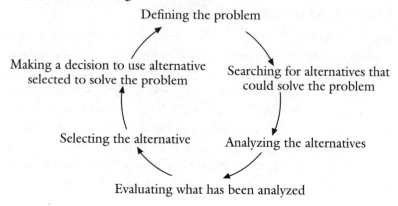

Defining the problem

Making a decision to use alternative selected to solve the problem

Searching for alternatives that could solve the problem

Selecting the alternative

Analyzing the alternatives

Evaluating what has been analyzed

(B) Strategic planning:

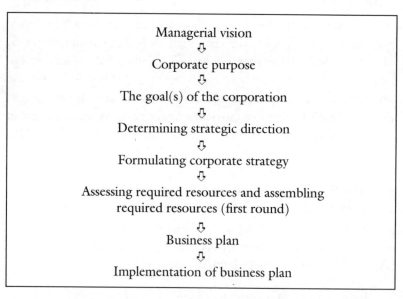

Managerial vision
⇩
Corporate purpose
⇩
The goal(s) of the corporation
⇩
Determining strategic direction
⇩
Formulating corporate strategy
⇩
Assessing required resources and assembling required resources (first round)
⇩
Business plan
⇩
Implementation of business plan

Honestly, can you say that there is something new in this definition, anything that has not been said and written about a million times before? Then, why is there so much excitement generated when this "old winter coat has only a dry cleaning"? Does it really make that much difference in knowledge, or does it really break new ground? The answer is, unfortunately, no. It has not provided us with any new knowledge, and certainly has not broken any new ground.

Aside from the definition, what is really involved in "Quality Management" is no mystery, and if anything can be accredited to it, it would be some old slogans such as: consumer (or customer) driven, system thinking, and leadership, all of which are so familiar to people in management positions, individuals who do research, and perhaps even laymen in the street. Figure 3.1 shows two examples which illustrate the application of system thinking, problem-solving in the first case and strategic planning in the other.

These figures are illustrative, and while in a practical situation, these might be variations from the scheme, the essence is constant. It has appeared in countless management or management related publications. What is so new about "system thinking"? Is this all that we can do for our business managers? Perhaps the challenge, if "Quality Management" is to be meaningful to managers, is to find another slogan or poster to convince managers to commit themselves to these glorious phrases such as "consumer driven", "system thinking" and "leadership" rather than providing only lip service.

4

The general management model: It gives our "B" schools some satisfaction, but what else?

Wisdom and fundamental truth can prevail in business management if, by working with people in harmony with the environment, businesses effectively, efficiently, and responsibly allocate resources for the purpose of creating wealth for the individual and adding value to society.

Introduction: The Purpose of "B" School Education, and the Search for an Identity for Business (Corporate) Management

The purpose of "B" school education

In the early 1960s, when a reputable institute of business and technology was just about to gain its university status, the author asked the President (then principal) the purpose of the Institute's business education. He looked puzzled for a while. Then as if talking to himself, he murmured: "Not bad, not bad at all. Our graduates get paid $600 a month." Some thirty years later, the author asked the same question to a well-known "B" school professor from a very reputable US university who was giving a seminar

74

on "B" school education to the staff members of a Pacific Rim university. Without hesitation, he responded: "to train our graduates to work for someone else's business, mainly, large corporations."

These two incidents reflect the reality that there has been no change in the minds of our business educators over a thirty-year period. The purpose of our "B" schools is to turn out individuals as employees of "entrepreneurless" corporate entities.

The search for an identity for corporate management

Some time in the late 1940s or early 1950s, when "B" schools were emerging in the university's mainstream from commerce programmes, quite a number of people were concerned about what was supposed to be taught in the business schools. For example, how does an MBA programme differ from an M.Comm degree or a BBA (Bachelor of Business Administration) or a B.Comm? It should be noted that although the MBA degree is offered by virtually all "B" schools, to this day, it is evident that the B.Comm and M.Comm degrees (with greater emphasis on economic analysis) are separated from "B" school programmes, and still very much alive in a large number of higher learning institutions throughout the world. Research findings concluded that "B" schools were not turning out individuals to be business leaders or leaders of society; instead, they were educating individuals to fill job vacancies. Awareness of the phenomenon led to a continuous surge of academics, writers and business managers, in general, searching for an identity. Many considerations arose including the possibility of forming a professional body to search for a philosophy appropriate to fit the model of corporate management. Conceptually, the wind has blown from a broader approach towards a search for a philosophy, modelled after John Maynard Keynes' "The General Theory of Employment Interest and Money", to the formulation of a General Theory of Management. Having failed in the above attempts, those who were concerned about the importance of the unification of management thoughts and practices that could be modelled by "B" school education, finally settled on a general framework of business management, that informally was labelled the General Management Model. Based on the "Theory of the Firm", the model assumes that a corporation is an entity whose objective is profit maximization. Expressed in the form of ROI through accounting manipulation, with the blessing of the stock exchange commissions, the model seems to have established itself in the higher learning institutions as the angel of business education. Fundamentally, the model is built on an assumption that a firm's objective is profit maximization, and profit goes to shareholders of the firm. Surely it is to their own satisfaction as well as to the delight of the stock exchanges, but are others connected with corporate endeavours equally satisfied?

The following section is an attempt to explore what has happened during the past four decades, in the jungle of management thought that led to the search for a general framework of business management in our "B" school education.

The Trouble with Considering Management as a Profession

The management profession's first pioneers operated in the factory system during the period of industrialization, in response to both the greater consumers' demand for goods and services and production pressure. Soon, the development of a more complicated market economy required better management of people, money and production and selling. Instead of going through a costly trial-and-error, business owners needed a more formalized system to learn some management techniques. Those who developed early scientific management thought, and the pioneers who were concerned with people made management a learned discipline.

Early in the twentieth century, it was perhaps at the University of Pennsylvania, US, that the first "B" school, the Wharton Business School, was created, thus paving the way for the further development of "B" schools today. It should be noted that the objective of the Wharton Business School was to train the younger generations of the rich for the purpose of taking care of their family businesses. It was in the post Second World War period that there was a great surge in business education, particularly in the US, and subsequently in Canada. It was also about the same time that universities and colleges themselves ventured into business education, that seemed to shift from the university tradition to develop broad liberal minded individuals to career-oriented applied management generalists. The development of these schools was to follow the examples of medicine, accountancy and a host of other "professions", in order to claim that "management" is a profession as well. "Professional management", so to speak, at times (even today) seemed to have penetrated many business educators' minds, and perhaps it was also for this reason that numerous studies took place with two notable ones, R. A. Gordon and J. E. Howell (1959), and F. C. Pierson (1959, p. 23). They concluded that "B" schools were not assuming their responsibilities to develop leaders in business, rather they were training individuals to fill job vacancies. To many, the conclusions of the two studies along with the work of others at that time turned business education in the academic community into a state of confusion and frustration and created uncertainty about what should be taught and how it should be taught. Thus, thoughts mushroomed about the possibility of developing management as a profession, much the same as other professions. Gordon and Howell were even more transparent about the matter of the so-called "professional management" or more appropriately about making management a profession. (See Table 4.1 and Figure 4.1.)

According to Gordon and Howell, first, the practice of a profession must rest on a systematic body of knowledge of substantial intellectual content and on the development of personal skills in the application of this knowledge to specific cases. Second, there must exist standards of professional conduct which take precedence over the goal of personal gain. They also stress that a profession has its association of members, among whose functions are the enforcement of standards, the advancement and dissemination of knowledge, and in the same degree, the control of entry into the profession. Finally, there is some prescribed way of entering the profession through the enforcement of minimum standards of training and competence (Gordon and Howell, 1959, pp. 69–70).

On the basis of the above, the authors concluded that in as much as there were efforts devoted to make business manage-

Table 4.1 Business (corporate) management needs: From war period (1940s) to near the end of the twentieth century

Driving force	Period	Business management needs
War (government) driven	The war period 1940s	Making war machines and supplies and meeting schedules in a captured market.
Production driven	1950s–1960s	Production (Engineering) efficiency. Transforming from war production to consumer market production.
Profit driven*	1970s–early 1980s	Profit squeeze. Corporations need trained financial control specialists to perform staff control function to improve profit.
Market driven	Mid-1980s–early 1990s	Market expansion, corporations need individuals who are capable of having a global vision, and managerial competence to expand operations beyond national boundaries.
Market driven, capital driven	Mid-1990s to the balance of 1990s	Global competition, rapid capital accumulation. Corporations need individuals to sustain and improve their global market position.
Environment protection and resources driven	Tail end of 1990 onwards	Rapid capital accumulation and global expansion exhausted investment opportunities. Corporations need creative individuals who could carry their corporation into the twenty-first century.

* Under the current situation, profit driven is a constant in corporate management, a factor of motivation.

ment into a body of knowledge, they failed to meet the other criteria. Therefore, based on their findings, they felt that business management was not a profession. It is fashionable these days, for people in any occupation to be considered professionals, including for example, computer consultants, pest controllers and real estate agents. Our democratic infrastructure allows these "professionals" to band together to freely form associations, drawing

Figure 4.1 The roundabout process of management education

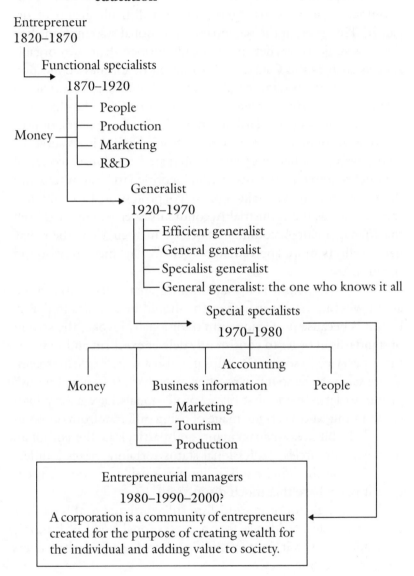

Entrepreneur
1820–1870

Functional specialists
1870–1920

Money
— People
— Production
— Marketing
— R&D

Generalist
1920–1970
— Efficient generalist
— General generalist
— Specialist generalist
— General generalist: the one who knows it all

Special specialists
1970–1980

Accounting

Money   Business information   People
— Marketing
— Tourism
— Production

Entrepreneurial managers
1980–1990–2000?
A corporation is a community of entrepreneurs
created for the purpose of creating wealth for
the individual and adding value to society.

members, elected officers, and set up admission standards. Unlike these groups, and perhaps similar to teachers in the academic environment, business managers can be a difficult group to organize. However, in the secondary educational system, Canadian business subjects teachers have already formed their own organizations such as the Canadian Association of Business Education Teachers. It has its own membership, and publishes a journal entitled *The Canadian Journal of Business Education*. They have not yet set up entrance standards as do professional engineers, or professional accountants. Presumably, a teacher of business subjects, who pays his or her dues and is obligated to observe codes of conduct required by the association is entitled to be admitted into the association. Incidentally, a branch of professional accountants, then known as the Industrial Accountants Society, renamed itself the Management Accountants Society. In their view, their new name reflects more appropriately their management function in the industry.

Why are academics and some business managers so concerned about whether or not management should be a profession? Perhaps it is because professionalism traditionally invokes the setting of standards in order to consistently deliver excellent and state-of-the-art service to society. Perhaps it is also a matter of self-esteem. There seems to be so much prestige given to "professional status" in our society. In the first place, professionals have a very high credit rating, and a certain image of being different from others in society. In business management, and particularly the corporate management circles, professional status certainly makes a difference. But on the other hand, if "professional" is so loosely used, does it really have that much distinction?

The struggle for professionalism still continues to this day, for there is no clear evidence that a professional body of business managers exists as it does in traditional professions such as engineering or accountancy.

The Challenge of Searching for a "Philosophy"

Accompanying the search for professionalism in management is the question of a managerial philosophy. Specifically, would it be possible to develop a philosophy for business management (or more correctly corporate management) that could be used to guide management in the marketplace?

Philosophy, as we know it, is the pursuit of wisdom, or a search for truth through logical reasoning. If corporate management is to have its own philosophy, it must demonstrate wisdom in the process of managing which reflects a fundamental truth that is meaningful to life. Unfortunately, corporate management practice under the influence of the theory of the firm has made it difficult to define a fitting philosophy for management as we know it today. The reasons are as follows: if there is to be a meaningful philosophy for corporate management, it must be reflected in the objectives of the corporation. To microeconomists, this objective is "profit maximization", and to accountants it is the "maximization of the rate of return on investment (ROI)". This concept is absolute as perceived by microeconomists, accountants and corporate equity investors, and possibly even corporate managers; unfortunately the concept of the maximization of the ROI does not necessarily make a good philosophy; after all, the desire for maximization is an innate biological instinct. There is no wisdom or fundamental truth by logical reasoning and no element of philosophy. If the maximization of behaviour in respect to ROI is to be considered a management philosophy, then any animal's desire for the maximization of food consumption would be a philosophy as well.

A philosophy must possess values that can be doctrinized as a fundamental truth which provides meaning to life. Religions and family systems are clear examples of such philosophies, but maximizing profit or ROI maximization endeavours are not in the same league. Countless profiteers or corporations have made their billions and were able to immortalize themselves by donating a fraction of their "profit" to the public domain and had their names engraved in gold in public places. Yet this still has nothing to do with philosophy, because it does not have the wisdom that reflects the fundamental truth in life.

Your values in life can preclude the necessity to accumulate financial wealth. For instance, religion has its own value system, more often than not a proud tradition that constitutes its believers' way of life. As an example, a member of a religious group won a lottery worth millions and millions of dollars, and out of the goodness of his heart, he decided to share his fortune with others and donated a portion of his prize money to his religious denomination. His donation, however, was rejected, because in their belief, gambling is forbidden. By purchasing a lottery ticket, he had violated the fundamental belief that you must not gamble. Mother Teresa who donated her Nobel Prize money to the cause of helping the poor and underprivileged is another case in point. It must also be emphasized that Mother Teresa's endeavours do not in any way involve the accumulation of personal financial wealth.

Wisdom and fundamental truth can prevail in business management as well, if by working with people and in cooperation with the environment, businesses effectively and efficiently allocate resources for the purpose of creating wealth for the individual and adding value to society.

Social Responsibility and Corporate Management

Academics also found themselves failing on two fronts: business management still could not legitimately be considered a profession, and academics had been unable to find a fit and a philosophy for management, which could accommodate the profit maximization objectives of the firm. Consequently, a large number of academics, including George Steiner and Frederick D. Sturtivant, made great attempts to redefine the corporation's goal to embrace social responsibilities. Their argument was that since the corporation as an entity is a citizen of the community, it ought to have a concern for the well-being of other citizens by incorporating goals such as concern for employment, water pollution abatement and control of industry waste (Steiner, 1975). This concept was particularly popular in the post Second World War period, and the thought was that the corporation by taking up those responsibilities, would be able to relieve public outcry about the monitoring of corporate power and lessened government intervention. Not everyone was agreeable to the thought of requiring corporations to take up social responsibilities, including the Nobel laureate Milton Friedman who advocated that to take on non-economic social goals would violate the interests of the stockholders, resulting in business imposing its views on others, and would be fundamentally subversive to the free enterprise economy (Friedman, 1962; see also Leavitt, 1958).

Some ten years later, in a book entitled *The Higher Taste* (International Society for Krishna Consciousness, 1983), it was noted:

> Because of the filthy, overcrowded conditions forced upon animals by the livestock industry, vast amounts of antibiotics must be used. But such rampant use of antibiotics naturally creates antibiotic-resistant bacteria that are passed on to those who eat the meat. The FDA (Federal Department of Agriculture) estimated that penicillin and tetracycline save the meat industry $1.9 billion a year, giving them sufficient reason to overlook the potential health hazards.

In the late 1980s, when the Chairman of Union Carbide first heard about the disaster in Bhopal, he stated that he would devote his life to making right what had gone so wrong for so many victims. But following the accident, Union Carbide proceeded to liquidate a substantial portion of its assets and gave them out to shareholders in special dividends, thus reducing the corporation's potential ability to give payouts to the victims. Investors who bought shares after the disaster tripled their money as billions were paid out to Wall Street speculators, institutions and arbitrageurs. In India, years after the accident, the majority of the 200,000 victims exposed to the deadly gas suffer corneal opacity or blurred vision (Hawken, 1994, p. 116). It was at about the same time:

> American executives at Shell Oil Co. in charge of manufacturing DBCP [a type of pesticide] were found not liable for the 1,000 Costa Rican employees of Standard Brands who became sterile after working with this chemical, nor was Standard Brands "liable" for shipping the remaining 45,000 gallons of DBCP inventory to Honduras after the pesticide was banned in Costa Rica (Hawken, 1994, p. 119).

Perhaps these are a few simple examples for what in the name of cost and profit, Friedman's free enterprise economy can do to people without corporations assuming their share of social responsibility.

Friedman's idea of the free enterprise economy without corporate social responsibility is, in the author's opinion, no less than an intellectual justification for colonialism, imperialism, poverty, wars, organized crime and social disorder. Indeed this reasoning is flawed by a fundamental paradox, for if the free enterprise economy is absolute, how can there be any free enterprise? The strong and powerful corporate giants would dominate the market, build up high entry barriers and eliminate competition and consumers' "democratically given right" of freedom of choice. It is interesting that Friedman also noted that: "Freedom is a tenable objective only for responsible individuals. We do not believe in freedom for a madman or children" (Friedman, 1962, p. 33). Would Friedman consider corporate executives whose decisions give no consideration to environmental health, or car makers who refuse to recall potentially life threatening vehicles responsible individuals or madmen?[1] Or perhaps the natural environment and the individual car and truck owners are not entitled the consideration for their "right" to life outside of the "capitalist" manifesto, and free enterprise economy? It must not be mistaken that the reference made to corporate social responsibility suggests government intervention. On the contrary, corporations (through their executives' action) by assuming greater social responsibility will reduce or eliminate government intervention. "Freedom is a tenable objective only for responsible individuals." The question is: How can corporate executives and managers be made more responsible? Perhaps, the starting point is to re-examine and redefine what the "B" school educational objectives should be.

Despite objections made by the powerful few, a sense of social responsibility has, for better or worse, penetrated into the corporate structure. The sad thing is that under the bottomline approach to corporate management, these social responsibilities are often no more than just after dinner talk, or more to the point, a public relations front with little or no substance.

The Attempt to Form a General Theory of Management

Confronted with the fruitless efforts to search for a philosophy for business management, scholars in the field of business management apparently felt that by consolidating various thoughts expressed by experts throughout the years, a general theory of management could be formed. This follows in the tradition of John Maynard Keynes' *The Theory of Employment Interest and Money* (1964), which is a collection of definitions and ideas, propensities and factors, spoken or written about in the major schools of thought directly related to the management of corporate entities. They include persons such as William Frederick who in his work *The Next Step in Management Science* (1963) viewed that, by drawing the best of all works of business management, you could form a body of management thought expressed in a logical consistency, to be put forth as a theory that can be applied to the process of management. Similarly, Arthur C. Laufer developed a taxonomic matrix that attempts to classify various management concepts in order to aid the development of a general theory. However, some authors felt that since management is such a complicated discipline which requires value judgements, such a theory could lack clearness and generality in its premises, and attempting to unify the thoughts and management descriptions would be fruitless. To this day, there has been no unification leading to the successful formulation of a general theory of management. Harold Koontz, in his article "The Management Theory Jungle" (1961), has grouped other existing works into six major categories:

1. *The management process school* – believes that management is a process of getting things done through and with people operating in organized groups.

2. *The human behaviour school* – believes that it concerns human relations, leadership and the use of behavioural science approach to business management.

3. *The empirical school* – believes that management is a study of experience, typically the use of case methods.

4. *The social system school* – perceives business management as a system of cultural interrelationships and interactions.

5. *The decision theory school* – focuses on the study of how decisions are made, and the selection of alternative courses of action.

6. *The mathematical school* – views business management as a model building exercise. It includes such disciplines as operations research and analysis. To a lesser extent, it is a study of managerial economics through decision-making mathematical expressions.

Of course, there have been other emerging contributions which attempt either to conceptualize or systematize the business management process, including the use of an interdisciplinary approach, comparative management and intramanagement approach to study the business management process.

From Informal Traditional Management Theory to a General Management Model Approach to Business Management Education

The jungle of management thought has not been able to provide a general theory of management acceptable to all concerned individuals, but some people have thought of focusing on more practical aspects of the process; the functional approach, and the problem-solving approach[2] which are two basic learning apparatus. The functional approach, as we know, amplifies the idea of planning, organizing, staffing, directing and controlling. Problem-solving may be typically described as: identifying the problem, searching for alternatives that can be used to solve the problem, evaluating the alternatives, and finally selecting an alternative that could be used to solve the problem. These traditional approaches may not be formally doctrinalized. They are to this day very much used in learning and teaching business management, even though critics are not happy about the lack of the fundamental management challenge with respect to the making of discretionary decisions.

Perhaps it was not by design, but by sheer necessity, that emerging from the general framework of business management is a model of general management through logical reasoning that could accommodate major thought patterns in the area. Although informal, our "B" schools worldwide, with some variation, tend to generally evolve, create or modify their curricula along the general management model idea, though without, as yet, a formal structure.

The General Management Model

In essence, the model follows much the same as "the bottomline" approach to business management. The bottomline is of course, profit, the maximization of the ROI, or the goal of the firm as advocated by microeconomists and used by accountants. Thus, it is apparent that the essence of corporate management, reflected in the general framework of corporate management (the general management model), is built around the financial success of the firm to meet the expectation of its shareholders.[3] Since the stock exchange is the money market where corporations fulfil their financial needs, it is in reality, the stock exchange that dominates the corporate managers' mindset, as well as the "B" schools' curricula. Therefore, it is not an overstatement to say that the stock exchanges are the power sources and the towers of financial giants and that they dominate "B" school education. Without accounting for them all other thoughts about business education may just as well be for the birds.

During the 1960s and 1970s there were sharp criticisms from the business community that "B" schools were not turning out graduates to meet the corporations' needs.

Today, the "B" schools seem to have a sense of satisfaction: "now we have finally, through the general management model touched the corporate decision makers' minds." A sense of satisfaction was clearly expressed through the buildings, and schools, renamed after financial donors who have made billions in the financial markets or through corporate expansion. More directly, the ultimate gratification for the corporate giants was to have them sit on the advisory boards or the board of governors of the university to tell or advise them on what to teach and what students need to learn. The governments, of course, are equally pleased, since it is fashionable for everything to be privatized (university reliance on corporate support is a form of privatization). Better still, as some Japanese companies have done in the US and other parts of the world, simply let corporations buy out universities and make universities function much in the same way as the corporations' own laboratories, and make them wholly owned corporate subsidiaries. This can certainly be heralded as a triumph, a triumph of private capitalism within Milton Friedman's free enterprise economy. Anyone attempting to oppose this reality would make a lot of powerful people raise their eyebrows: "How dare you go against the motherhood of capitalism and a free enterprise economy!"

What is a general management model? There is no typical model that can be used to illustrate the thought. However the general idea of the model is very much like Figure 4.2.

It should be noted that a number of engineering based "B" schools may have a slightly different approach to business management.

Specifically, they tend to emphasize more the use of a "cost efficient" resources utilization approach; however, they are unable to escape the reality of "bottomline" accounting dominated measurement.

There is a question of social responsibility and entrepreneurship. Under the bottomline approach of business management, entrepreneurship is not a factor. According to microeconomists, accountants, stock exchanges and the legal profession, a corporation is an entity, and everyone working in the entity is an employee. In essence, a corporation is entrepreneurless, and how to fit it into the scheme is anyone's guess. In large corporate structures, staff control is the essence of corporate management, where staff control relies on corporate by-laws, rules, regulations, procedures and manuals. Hence, it would be difficult to visualize how a management emphasis can be "entrepreneurial", if innovation and creativeness are not in the books of rules, regulations, procedures and manuals. Moreover, to this day, there is a misconception that people tend to view "entrepreneurship" to be about starting your own business. Although some universities do offer entrepreneurship in their curricula, nevertheless it is under the auspices of the general management model.

Figure 4.2   An illustration of a general management model

"B" school graduates career path	Working for someone else's business, primarily, large corporations
Focusing on	Financial success
Measured by	Accounting method
Monitored by	Stock exchange commissions in conjunction with recognized professional accounting bodies
Expressed in the form of profit based on	Maximizing the rate of return on shareholders' investment; bottomline approach notion of business management

Curriculum based on the need to comply with bottomline approach notion of business management:

The core	1. Understanding how business and corporations work in the marketplace, the BASICS:
	• Introduction to business or business management
	• Economics, particularly micro theory
	• Organizational behaviour
	2. TOOLS
	• Accounting
	• Statistics
	• Business communication
	• Computerized business information systems
	• Research methodology
The functions	3. FUNCTIONS WITH A CORPORATE STRUCTURE
	• Marketing
	• Production/operations
	• Human relations/personnel
	• Finance/corporate financing
	• Information/communication
	• International business

Figure 4.2 (cont'd)

The management	4.	**CORPORATE MANAGEMENT**

- Management theories
- Decision theories
- Corporate strategy/strategic management
- Marketing management
- Operations management
- Information management
- Technology management (not yet a popular subject)
- Quality management
- Human resources management
- Financial management
- Managerial accounting
- Managerial economics
- Operational research and analysis
- Management control systems
- Small business management
- Government relations
- Entrepreneurship and/or new venture creation

The advancement	5.	**ADVANCED SUBJECTS**

- More in-depth subjects in any of the above areas
- Subjects will lead into a specialization and/or the executive MBA programme. Recently, Master of Professional Accounting (M.Prof.Acc.) and other Master's degree affiliated with business management.

The specialization*	6.	**SPECIALIZED SUBJECT IN ANY OF THE FOLLOWING MANAGEMENT AREAS**

- Marketing
- Accounting: affiliated with and leading to an accounting professional designation

Figure 4.2 (cont'd)

- Financial management
- System analysis
- Other specializations

| Schools of thought that service business school education: | The traditional views that see management as practical functions: |

A. The functions of:

- The management process school
- The human behaviour school
- The social system school
- The decision theory school
- The mathematical school

- Planning
- Organizing
- Staffing
- Directing
- Controlling

B. The function of problem-solving:

- Identifying the problem
- Searching for alternatives to solve the problem
- Evaluating alternatives
- Selecting the alternative to solve the problem

---

* It is assumed that once an individual entered a specialized area, his or her career path would either be working in someone else's business or entering a profession such as accountancy.

Note: The above model, in essence, is how the author perceives what the general management model would be. For the purpose of preparing the above, the author had the opportunity to view approximately 500 "B" schools' curricula, reviewed available literature and discussed the matter with a number of academic colleagues. Unfortunately, in as much as most of us agree that there is a form of general management model as described in management conferences or seminars, it is not possible to trace the origin or locate a model that has been fully developed. For this reason, the author wishes to acknowledge the presence of many unknown contributors in the field of management discipline and takes full responsibility for any irregularities that may occur in the development of the model presented here.

In contrast, social responsibilities are commonly acknowledged as part of corporate consideration. It consists normally of responsibilities to consumers, community and employees as follows:

To consumers:

1. Reasonable price, competitive but not necessary the lowest.

2. Quality product; to support this, they tend to have a return policy and goods satisfaction or money back guarantees.

3. Responsible advertising to ensure that all ad materials are not misleading.

To the community:

1. To support education; donations and corporate sponsorship are seen all over educational institutions.

2. To support the art and community projects; all you have to do is to look at art galleries or other cultural events, and see how many corporate names you can find.

3. To support medical research and health care.

4. To support the environment protection; any corporation spending money on conservation shouts it out as loud as they can.

To employees:

1. The usual things you can find in any personal or industrial relations texts: fair wages, good benefits, job security (the best is expressed by followers of Theory Z) and job satisfaction.

There is no doubt, to the disappointment of Milton Friedman, that with or without the free enterprise economy, social responsibilities have made their place albeit minimally in the corporate management manual. Unfortunately, when it comes to the bottom-line approach to corporate management, all these social concerns become little more than topics for idle conversation at a tea break. The argument is: "If you are dead, how can you reach the promised land? Profit, maximizing ROI, the first, and the last will always be the bottomline of corporate management. Amen."

The general management model, for all its worth, has answered the business community's concern for not being able to turn out graduates to meet their needs. What more do corporate executives expect? Under the circumstances, the chances of anyone criticizing the wisdom of using the bottomline approach to corporate management is perhaps slimmer than winning the "Loto 649" lottery.

Sure enough, in appreciation of what the "B" schools have done, there are corporate donations (mostly, for tax write-offs), and corporate executives sitting on the advisory or governing boards telling "B" schools what should be taught and what MBAs should learn. All of these must have given the "B" schools a great sense of satisfaction. Then some may say: "Who cares!" Yet, there are a lot of people with plenty of questions. For a start: "If our "B" schools only turn out individuals to be employees for large corporations, do they still have the same entrepreneurial drive as entrepreneurs who control their own destinies?"

The Tragedy

As the whole world celebrates corporate success through expansion, merger acquisition and globalization, it is apparent everyone believes that the "free market" driven by "profit motive" is the most appropriate mechanism for economic growth. Consequently, our universities, and particularly the "B" schools behave in much the same way as the corporate management, driven by the need for "funds", and apparently listening to everything that corporations have to say. This includes curriculum development, the use of case material for teaching, research efforts, staff recruitment and promotion, all of which seems to have been based on what the "business" wants. This has been seen as being practical and relevant. New thinking, and new thoughts are considered to be impractical, useless and overly theoretical. The tragedy may not be readily apparent, but we have made our best qualified managers behave in an animal-like fashion: striving for short-term performance and instant success at the cost of the people and the environment. Corporations support what they believe to be good educational institutions particularly the ones which turn out individuals to meet the corporations' needs, and give them financial support through contributions, providing a good image, on the one hand, and gaining benefit from tax breaks on the other.

Some corporations are not particularly satisfied with what the universities can do for them and have learned from the Japanese

(Toyota, for example) to create their own universities (such as Motorola). If the trend continues, there will be no academic free thinking and everything would go by the book, corporate manual and procedures. Then we would only teach our best people to follow manuals, procedures and systems, and there would be no thinkers or perhaps, entrepreneurs. Perhaps we should turn back to the apprentice system (see Figure 4.3), and let the universities be universities, even without funds, then let them be, and they may end up starving to death. If the days come as described, would there be any greater tragedy? Then Information Technology (IT) experts might say: "Who needs universities? Self-motivated individuals can always learn from the Internet." But can software marketers offer MBA and/or Ph.D. degrees through the Internet?

A Model for Entrepreneurship Education

While some individuals in business have a notion that entrepreneurship cannot be taught, a large number of people believe that even if this is so, students can learn entrepreneurial skills. The entrepreneurial mindset will be dealt with in Chapter 12. For the purpose of entrepreneurship education, it is argued that entrepreneurship can be taught and should be taught, but it is necessary to appreciate that entrepreneurship should not be limited to the traditional notion only applicable to venture founders and those owning their businesses. On the basis of this, a suggested "Entrepreneurship" educational model is illustrated in Figure 4.4.

Figure 4.3   The transformation of "B" school education

Apprentice system (corporations training their own employees based on corporate needs). No need for "B" schools to do the job.

The birth of "B" school, training individuals to work for someone else's business (funded by tuition, government grants).

"B" school generates some part of its revenue by being involved in business attachment, doing projects for business, providing consultations to business, and relying in part, on corporate contribution.

"B" school turned to be "fund driven" institutions, relying on corporate support heavily. Curriculum development and teaching slanted towards practical knowledge, ready-made materials, the use of case studies, and follow what corporations do.

Corporations are not particularly happy; "B" school provides more liberal thinking type of environment. Corporations need to make more profit, and "B" school graduates cannot deliver.

Forget about "B" school, let us (corporations) have our own colleges and universities to turn out true blue blood corporate men and women.

Back to the old days and make management apprenticed MBAs. Amen!

Figure 4.4   A suggested "entrepreneurship" educational model

Based on the broad definition: entrepreneurship is doing something new (creative) and/or something different (innovative) for the purpose of creating wealth for the individual and adding value to society.

Mindset

To develop an individual's understanding that we are energy ourselves. As energy we are naturally creative, and entrepreneurship is a way of life. Economics is not for a few to make profit but for directing and allocating limited resources in different combinations to increase the utility function for humans and other living beings' consumption.

Knowledge

Entrepreneurship is a set of knowledge. It has the methodology (empirical, descriptive) of research and learning, and is a discipline in its own right. It is applicable to all human endeavour, and has a science origin.

Subset: For Learning and Research

- Entrepreneurship Theory
- Corporation Entrepreneurship
- Technical Entrepreneurship
- Family Business
- Entrepreneurship in the Public Sector
- Women Entrepreneurship
- Parenting Entrepreneurship
- Entrepreneurial Skills
- Entrepreneurial Attributes and their Development
- Others

Application

All subset in the knowledge of entrepreneurship can be applied in areas such as:

- New Venture Creation Strategy
- Entrepreneurial Approach to Corporate Management
- Motivating Employees and Developing a Corporation into a Community of Entrepreneurs
- Ownership Participation and Stakeholder's Concept Development
- Value Added Approach to Corporate Financial Reporting

5

The cost of capitalism

We fear death as it ends life, but by continuing blindly to manage and expand corporations based on currently perceived capitalistic scheme of expansion, the consequence will be even more frightening than death, because it will be the "end of birth".

Introduction: The Most Dangerous Animal in the World – Calgary Zoo, Canada

The Calgary Zoo administration posted a sign on the side of a major road leading to the Eurasia exhibition area saying: "The Most Dangerous Animal in the World." At first glance, you would look for an exhibit cage containing this "most dangerous creature", but soon the visitor realizes there is no creature. Instead there is a free-standing iron fence erected behind the sign in front of which people can take photos of themselves. It is meant to be a sardonic joke, a bitter reflection of reality, at least from the non-human animal's point of view. We kill and destroy animals, the environment, and more often than not, ourselves as well. We do this not by using our physical strength (like animals), but with our intellect as well – we have our killing planned well into the future. Moreover, the destruction of our planet has gone on unnoticed for much of history. The destruction is unnoticed, because both our planet and the environment have been silent victims of "capitalistic growth". What capitalism wants it takes and the earth has had no voice to protest. Perhaps the Calgary Zoo administration is right: We are the "most dangerous animal in the world".

From Left to Right

As human beings struggle for survival, we seek ways and means to better ourselves in terms of materialistic, physical and psychological needs and satisfaction. Less than a century ago, people in Russia, Europe, Asia, Africa and even in America followed the untested Marxist-Leninist doctrine. An honest attempt was made to socialize production and distribution to relieve people of poverty and capitalistic domination – an ideology developed as an offspring of classic economics. The fall of the Berlin Wall and the collapse of the Soviet empire pushed everyone onto the lap of capitalism – communism's 200-year-old brother. Political leaders, business executives and, of course, some of our academic friends precipitously fell in love with the idea that only corporate expansion would bail us out of all our miseries. Countries previously allied with the former Soviet Union have all swung to the right.[1] Even China, Vietnam and Cuba, who still nominally remain "communist", practise market economies within a socialist governance. For all its worth, the adoption of capitalistic scheme of expansion seems to have penetrated our minds as the solution that will allow most of us to have a decent standard of living. It appears that the triumph of "capitalism" has finally arrived in "reverse" at the tail-end of the twentieth century.

The Emperor's New Clothes

No matter how much we talk about the information highway, it can still only provide us with the information that we put into it, and thus will never inform us of the true cost of capitalism to our society. The politicians' typical term of office is four to five years, and including re-election for another term or two, seldom results in a stay in office of more than twelve years. It would be unlikely for them to tell people about the true cost of capitalism. In rare cases, they know the true cost of capitalism to humanity, and question the wisdom of the happy and convenient tripod marriage between free enterprise, capitalism and democracy; rare still, some may openly challenge corporate power. For example, in 1962, the late US President J. F. Kennedy actually made US Steel back off from a deal of raising steel prices. He had no legal authority (as some free enterprise advocates would claim) to invoke such legislation, but Kennedy publicly painted the steel barons as evil profiteers (*Newsweek*, February 1995, p. 25). Yet this example is the exception rather than the rule.

In fact most of us know little about capitalism, and we often have blind faith in what capitalism has done for this world. After all, we do enjoy the harvest of franchised capitalism. Of course, there are also people who may simply be unwilling to admit their ignorance, and still others who have learned to live well and trouble free by acting and behaving as the good citizens in the "Emperor's New Clothes" story, going along with the public and watching the emperor parade through the street wearing his "birthday suit". Rather than telling the truth that the Emperor is naked parading on the road, they join the others praising the emperor's "birthday suit": "Wow! What a beautiful robe."

Can You Put Out the Forest Fire with a Garden Hose?

On the business side and in the marketplace, almost without exception, our corporate managers are constantly pressured by short-term performance (quarterly, semi-annually or annually, or perhaps with a little luck three to five years). They have the mentality that "life is too short to worry about the future" and "you cannot reach the promised land if you are dead". Even though it is possible for them to improve revenue, to work out a true cost and care about our planet and protect the environment it is too long a commitment. Under close monitoring by staff in the organization, how is it possible for managers to talk about the planet and the environment before profit? Moreover, although professional accountants are "toying" with the possibility of initiating some kind of accounting for social responsibility, who knows how long it will take, given that there has not been much change in accounting since centuries ago the Italian monks invented debit and credit for bookkeeping. True enough, there have been some candid efforts to develop accounting postulates or principles to practise, but to consider the cost of capitalism to humanity and incorporate it into the cost structure of a corporate operation before deriving a profit is absolutely unheard of. Perhaps out of ignorance, or unwillingness, neither have economists included ecological considerations into models of econometrics. With accountants maintaining the status quo, it is unlikely that they will go along with economists and ecologists and together work out a sensible mechanism through business management to slow down the slaughter of our planet and deterioration of the environment.

Without a doubt, concern for the environment and caring for our planet have caught public attention. Environmentalists and governments have taken some initiatives, and some universities and the business community offer courses on caring for the environment. Examples are the creation of the US National Commission for the Environment (by a group of influential people) and the impressive Business Council on Substantiable Development led by billionaire industrialist Stephen Schmidheiny with members on the Council representing the world's largest and most powerful organizations. But these are not cause for celebration, because corporations do not accept anything that will put the environment before profit. The oil companies lobbied, delayed and sandbagged the Environmental Protection Agency's Clean Air Act of 1970, and attempted to weaken the legislation for its revision in 1990 for reasons that a stronger provision for environmental safeguards would cost shareholders more money. Mitsubishi of Japan managed to shelve a United Nations agency's effort to impose mandatory rules regulating the conduct of corporations with respect to the environment, replacing them with a voluntary code of conduct drawn up by the corporations themselves (Hawken, 1994, pp. 111, 168). Then there is the question of corporate behaviour unreachable by law. For example, exporting plastic waste and other garbage is legal in the US. Corporations which have garbage problems are giving a few dollars to countries who, apparently in need of foreign exchange would help solve the woes. China was one such country. But China in recent years has its own anti-pollution law which prohibits the import of garbage, so the garbage was left in Hong Kong instead. What is disturbing is a punch line frequently used by American political leaders: "We have our differences and difficulties, but when in crisis, we Americans, stick together to deal with the crisis." Why can't corporate executives realize that although "we may have different political systems and religious beliefs in different countries, garbage pollution is a world-wide crisis. Shouldn't we deal with the crisis together, rather than dump our waste on our neighbours and telling them: 'Here is a few dollars, you can have our garbage'"? These and other countless reported and unreported happenings around the world illustrate that assuming corporate responsibilities without understanding their nature is like attempting to put out a forest fire with a garden hose in the corporations' bid to protect the environment.

Only Corporate Management can Prevent our Course of Environmental Destruction Plotted by Capitalism

Despite the fact that environmentalists and government efforts have impacted the environmental policies of the corporations to some extent, real action must still come from the business sector. What is needed is a change in the mindset of corporate managers and a commitment to care for our earth. To do so, it is necessary to change our views on how we practise capitalism, what it does for us and what the cost is. Are there better options to satisfy human needs and wants without causing so much harm to our people, and planet?

In 1992 the author was given the responsibility by the International Council for Small Business (ICSB) to organize and chair its 37th World Conference on Small Business and Entrepreneurship Development. The Conference theme, "Enterprising in Partnership with the Environment", was dedicated to the business community and environment. There was nothing new about the topic, since prior to the Industrial Revolution, the awareness of our surroundings and environment was often much greater. The Industrial Revolution has done a lot for us, but it also set the stage for human beings to take indiscriminately from the environment, both for our needs and our pleasure, much as we sometimes kill animals in captivity for recreation.

In the session on entrepreneurship and the environment, the discussion touched on every issue imaginable, including the relationship between profit and caring for the environment. What should the government do? How can environmentally conscious corporations at home compete with those in some other nations where environmental care is not an issue? No perfect answers were found to these questions and many others, but the participants agreed in general that there are two fundamental causes to the deterioration of the environment. One cause is an expanding population problem which the session leader decided not to discuss, while the other is business. When asked by the session leader to conclude the session, the author ended with the following:

> It is possible to make a parallel between eating and health, and business and humanity. In the first instance, we need food to survive, but excessive food is also the cause of health deterioration and death. Most foods taken in excess are unhealthy, but if we don't eat, we'll die. We cannot escape death, but by eating sensibly and reasonably, we can live longer, healthier and even happier. An analogous relationship exists between business and everyday life in the modern society. Without business, there can be no exchange system. With no exchange system, we don't get what we need to survive (of course, some of us could always live as other animals, as part of a natural ecosystem). However, the end result of aimless capitalistic expansion (too much business so-to-speak) and continuously pushing short term results will be the destruction of the environment. Eventually society as we know it will end. The challenge is for our corporate managers to commit themselves to managing business responsibly as true entrepreneurial companies with responsibility not only to corporate shareholders, but to all people and the environment as well. There is a saying: If I give man a fish, then I feed him for a day. If I teach a man how to fish, then I feed him for life. But if I teach a man to fish responsibly, then not only do I feed him but the rest of his village as well as for generations to come (Kao, 1995, p. xiii).

Poverty is not the Enemy of the Environment; Blindly Expanding Capitalism is

> For the poor, anything is reusable, to the rich, everything is a throw away. Who is the enemy of the environment?

In the report by the National Commission on the Environment entitled *Choosing a Sustainable Future*, the authors (Commission members, including at least two corporate executives) remarked: "Poverty is the enemy of the environment. Environmental protection is not possible where poverty is pervasive and the quality of life degraded" (1993, p. xv). The Commission members were of a high calibre and included notable figures such as Paul H. O'Neill, Chairman of the Board and Chief Executive Officer of Alcoa, John Bryson, Chairman and Chief Executive Officer of Southern California Edison Co., Steven C. Rockefeller, Professor of Religion, and Cruz A. Matos, Environmental Consultant, former Chief Technical Advisor, United Nations. What was surprising was that none of the members realized that although poverty is the enemy of humanity, it is capitalist perception and practice that are the enemy of the environment.

Human beings fear death because it ends life, but if we continue blindly to manage and expand corporations based on what is currently perceived as the capitalistic ideology, the consequences will be even more frightening than death, because it could be the end of birth.

An Illustration of the Meaning of the End of Birth

Green Island is part of the South Pacific, and situated off a small nation. As its name implies, it is one of nature's gardens. It has a rainforest ecosystem and has an extremely plush and varied flora; its fauna includes hundreds of rare birds and small mammal species. Several inlets and secluded coves provide a peaceful and natural environment for many fish and other water species including highly prized jumbo-size shrimp. This environment attracted Joe, a large investor who already owned a fishery and food processing empire. At first, Joe was happy simply fishing for shrimp around the island, but he soon saw a window of opportunity in Green Island and visualized that he could turn it into a large shrimp farm. This would be a profitable venture, since he already possessed the technology and all he needed was the right to proceed with another of his conquests. Joe would be, as put in some academic writings, a "blue blooded" entrepreneur. Joe managed to acquire a ninety-nine-year lease from the local government, and proceeded with his venture some seventeen years ago (1980). The conversion of this island to a shrimp farm took only eight months. First, most of the trees were removed from extensive tracts of land, and a large number of shrimp farm patches were constructed, similar to rice fields along the banks of the waterways, pumping ocean water onto the patches. In no time at all, production yield mounted, and by the end of 1985, the operation began to show a sizeable profit with increases for four years. Unfortunately, by the end of 1989, the production yield suddenly dropped, and coincidentally with an alarming death rate in other animal populations. Tests determined that this mortality was due to excessive accumulation of salt caused by the repeated replenishing of salt water into the farm. Nothing could be done to remove the salt, and with time, large amounts of salt penetrated into the surrounding soil, which killed the remaining trees and anything green left on the island. By 1994, the entire shrimp farm existed on a heavy concentration of salt. The astonishing assault on the island's ecosystem by the accumulation of salt resulted in zero growth among the trees, plants and animals. This island's nightmare is perhaps an extreme example, but it illustrates how business not only causes death, but also ends birth.

What happened in Green Island is not just an isolated incident, but can be described as calculated greed under capitalism. Joe could have profited from simply shrimp fishing rather than farming. Albeit fishing would have resulted in lower profits all other living beings in the island could have been saved, and with care for the environment, the ecosystem could have survived indefinitely. This is a striking example illustrating that it is man's choice on how to live, either cooperating and benefiting from nature, or simply conquering and destroying it.

Capitalistic Expansion, Corporate Value and Tradition

By geographical size, Japan is a small nation in comparison with its neighbours, China and Russia, or with many other countries across the Pacific and around the world. She has no significant natural resources, and in particular, lacks petroleum reserves. However, Japan is one of the most commercially powerful nations in the world, and her economic performance is so well-known that not a single person in the world can deny this reality. It is estimated that more than 10,000 Japanese corporations now operate on a worldwide basis. They include operations in electronics, car manufacturing, supermarkets, chemical companies, communications, media and the entertainment industry, banks and department stores which spread out in virtually all major cities of the world. It was estimated that in 1993, at least 2,000 corporations operated in the US, and about one half of this amount have established their foothold in Europe. Moreover, Japan's capitalistic expansion has also included the acquisition of research laboratories, colleges and universities in the US, England and other countries. Mergers, acquisitions and buy-outs have all been part of Japan's capitalistic expansion strategy, particularly in the US. For the record, Sony took over CBS records, and JVC, a branch of Mitsubishi electronic corporation, invested $100 million in a US film studio. With a different strategy for expanding corporate operations in Europe, England has been used as a springboard for Japan to extend its arms all over Europe. Corporations involved in capitalistic expansion include car makers such as Toyota, Nissan and Honda, and electronics giants such as Sony, Hitachi, Toshiba and JVC. In recent years, expansion has also extended to China, Russia, and Eastern European countries; the fingers and toes of these same companies have also expanded into Central and South America. It goes without saying that there have been huge overseas capital investments in all parts of the world, most recently in developing countries with a need for capital to activate their own human and other forms of resources. Japanese corporate expansion seems to be a logical solution for these countries, while Japan, already having so much money in hand, continues to find investment opportunities where there are cheap labour costs as well as abundant resources. It is the same strategy behind the invasion of China in the 1930s and early 1940s. China's labour costs were cheap, particularly in the northeast region of the country, and underground resources were abundant. The only difference between then and now is the difference between the use of guns and the yen.

The Japanese capitalistic expansion overseas is not without repercussions, particularly in these developing countries. These countries need to develop their own industrial strategies to solve (1) employment problems; and (2) a shortage of foreign exchange both to balance their trade deficit and acquire hard currency to buy what their countries need (perhaps for the luxuries of a few, or to meet consumers' needs). They welcome Japanese capitalistic expansion as a solution to these problems, and in some way, it is considered to be an achievement; it indicates their country to be worthy of Japan's investment. It is just like having one stone killing "three" birds: (1) boosts the morale of the people; (2) heightens the status of the government; and (3) develops the nation's economic needs. But the honeymoon is not all that sweet. Japanese corporations are not so generous in their dealings with these countries. In the first place, factories are little more than assembly plants. Major decisions are made in Japan; in countries where the plants are located, a colonial type of operation exists with which many more Western nations are familiar, such as the branch plant operations experienced by Canadians in the early 1960s, where managers were not developed as top decision-makers. Virtually no opportunities exist for subsidiaries or branch plants to carry out their needed research. Above all, there has been very little technology transfer. Technology to the Japanese (perhaps to others as well) is like the operation of a submarine; it must be fast, silent and deep. A statement made by the Malaysian Prime Minister Dr Mahathir Mohamad is very much to the point. He said: "I ask the Japanese to look not only at what they can take, but what they can give.... we cannot, and will not remain merely known as hewers of wood and drawers of water" (*Asia Magazine*, 26–28 March 1993). Ironically, developing countries are in need of Japanese investment to develop their industry, and Japanese subsidiaries help with employment, but all these are not without cost. The profit earned by Japanese investors and corporations is either from its utilization of foreign cheap labour or natural resources. The situation is very much as it was in the 1960s and 1970s, when the oil producing nations complained about Western nations living luxuriously off cheap petroleum prices.

An analysis of the benefits of capitalistic expansion to Japan is of interest. The expansions have done a great deal of good for corporations such as Mitsubishi whose annual sales exceed the gross national product (GNP) of all but twelve countries globally (Marshall, 1991, p. 78), but they have not necessarily been good for the people in Japan. The most significant development is that low-skilled positions are being farmed out overseas, and robotics are being developed. Consequently, the legendary lifetime employment, and the big family concept which has been the foundation of Japanese industrial success appears to be a thing of the past. Japanese workers at home are being laid off just the same as in Western nations (Marshall, 1991, p. 78). This obviously is part of the cost of capitalistic expansion.

No Mercy for the Rainforests

The world's forests are disappearing at a rate of 15 million hectares each year, with most of the losses occurring in the rainforests of Africa, Asia and Latin America. Deforestation is just as old as human history, and people throughout the world have chopped down trees for a large number of reasons. Recently, however, deforestation has increased dramatically causing great concern.

About a decade ago, when the author was at Lake Geneva in Switzerland, he noticed a sign outside a McDonald's hamburger outlet: "Last year we served the world more than one billion hamburgers." Obviously, the people behind this must be proud of their achievement. If you were to pause and think for a minute, this represents not just an overwhelming number of cows needed to be slaughtered but also millions of acres of pasture land to be used. Consequently, bulldozers and chain saws are used to clear out forests in tropical places like Latin America and Southeast Asia. Even in a country such as Costa Rica whose government is known to be dedicated to the protection of its environment, massive cattle raising is favoured because it is a cheaper form of generating export revenue. Costa Rica has also shown significant deterioration of its forests. As described by the country's only astronaut Senor Franklin Chan, some twenty years ago, 75% of the country was visibly covered by green; but today, at best it is about 50%.[2]

In Asia, Japan is a heavy user of timber; almost one-third of it is hardwoods. Virtually all of it comes from Malaysia and Thailand, with Malaysia being the largest supplier. The giant logging companies, with the desire for foreign exchange, operate mercilessly, cutting down trees faster than the trees can grow (Marshall, 1991, pp. 120–21). Deforestation in Canada and the US is much the same as in Southeast Asian and Latin American nations, with the added problem that it takes much longer for a forest to mature in a cold climate than in the tropics.

In African countries, wooded areas are heavily depleted. In 1940, 40% of Ethiopia was covered by forest, but today, only 4% of the green land is left. It is estimated that if the situation continues, within a period of 40 to 50 years, there will only be a few forests left on the planet.

The use of timber for building structures and other things is considerable, but the most discouraging fact is the excessive amount of paper used for newsprint, even though most newsprint today is made from recycled paper. Bear in mind that recycled paper uses a whitening process which pollutes the environment. To make matters worse, the so-called newspapers printed on newsprint often contain little "news"; in some cases, over 80% consists of ads. Why do the daily newspapers use so much paper for so little real information? The answer is obvious; large corporations use every means available to get their message to consumers, the more numerous and larger the ads, the more effective they will be. The people who are responsible for the massive level of advertising are, of course, the publishers of the newspapers themselves, egged on by the agencies and corporations. They seem to have no idea how many trees have to be cut down before 180 pages (approximately 70 of one daily newspaper in Chicago) of ads can be printed. There is no concern from the corporation, the ad agencies, and the newspapers of course, who live out the idea: "In business, everything is possible, the only difference is price." Since the agencies and corporations are willing to pay, why should they care about trees, the environment and humanity? In one ad, one company promotes its product through an ad mindlessly and shamelessly saying: "We bring the future to you for your immediate enjoyment." It is almost like taking essential food away from an infant to fatten an already grossly inflated belly just for the sake of making the big bigger. What is most amazing is that virtually every single individual in one aspect, wishes to have children who will bear the family name and continue to live and prosper as we do, and naturally humanity as a whole will continue. But it is a known fact that a large number of individuals will do whatever they can to take away what belongs to our children for their immediate enjoyment, as if future human beings can live without life support. Unfortunately, the environmentalists' outcry has not been effective to convince business leaders to place environmental health before short-term profit. Also, religious leaders have not been able to do much to convince the extremely wealthy and rich to slow down their "greed". It is even worse that our political leaders, academic friends who advocate capitalism as the survivor of humanity have done nothing less than help spread the "maximizing profit" as the "corporate culture", without the knowledge and accounting for the true "full cost", or the cost of capitalism. And as long as we are unaware and unwilling to explore, accept and account for the "true full cost" of capitalism, and continue to advocate capitalistic expansion as our philosophical and academic ideal, we will continue to exploit and destroy our environment.

Corporate managers and investors must change their mindset, to realize that although life is limited by nature, we have no right to take what belongs to the children of the future for our present enjoyment.

The use of trees for timber, newsprint and packaging, and the elimination of forests for massive cattle-raising are the major causes of the destruction of plant and animal species, resulting from capitalistic expansion. If the present rates of habitat destruction continue, then some 40,000 to 60,000 vascular plants will face extinction within fifty years. The consequence of the continuing destruction of plants will threaten the world's food supplies, and undermine medical advances.

Tropical forest originally covered about 16 million square kilometres of the earth's land surface; today probably less than half remains. Although the tropical forest (including rainforest) accounts for only one-third of the world's total forest, it contains four-fifths of the world's vegetation: a single hectare of primary forest may support plant material weighing anywhere from 300 to 500 tonnes. More than that, it provides a habitat for about fifty per cent of all animal and bird species. As the tropical forest is destroyed, plant and animal life will disappear with it. Destruction of the tropical forest is not only limited to industrial and agricultural land, the tourism industry is just as destructive. Even environmental friendly nations such as Costa Rica cannot escape the temptation of tourist dollars. Capitalism in tourism development may be contrary to the country's overall "goodwill" in its concern for the environment, but people are urged to make sacrifices as seen from statements such as the following:

> The measures taken to recognize the project are stern and will demand sacrifices that all involved parties will have to bear with understanding. But because of this, Costa Rican tourism will be strengthened at the same time that it will point to new ways to follow, without which the whole concept of sustainability would lose its meaning.[2]

Unfortunately, the sacrifices are not merely made by people, but by the environment.

Industrial Waste, Fossil Fuel Pollution and the Car Culture

A publication entitled *The Corporate Way* released the following figures related to toxins released from US industrial plants (parent company 1987) with the following (Seager, 1990, p. 115):

- Toxic chemicals (thousands of tonnes):
  Aluminum Co. of America   452
  National Steel   349
  Du Pont de Nemours   154

British Petroleum	152
Monsanto	108

- Known cancer-causing chemicals (tonnes):

Pfizer	4,639
Eastman Kodak	4,538
Du Pont de Nemours	4,391
General Electric	3,775
Eli Lilly	3,277

Fossil fuels are oil, gas and coal. Most capital driven industrial economies are entirely fossil-fuel dependent. The price to pay (the cost of capitalism) of this dependency is severe pollution and the looming possibility of global warming. The burning of fossil fuels produces a number of pollutants, including sulphur dioxide, nitrogen oxides, particulates and carbon monoxide. The most serious fossil fuel pollutant is carbon dioxide. This is not in itself toxic, but in conjunction with others is in the primary greenhouse gas. Worldwide releases of carbon dioxide from fossil fuel combustion was estimated in the late 1980s as amounting to 2 billion tonnes per year.

Attempts to curb the problem are being thwarted by countries most responsible. In the autumn of 1989, sixty-five countries sponsored an international resolution to cut carbon dioxide emissions in industrial countries. It was defeated by objections from Japan, the US and the then USSR (Hawken, 1994).

Oil Pollution and Car Makers

The lifestyle of the industrialized countries is entirely dependent on oil. Oil is a big business, and the oil industry as a whole along with car makers virtually dominates the entire world economy. For anyone attempting to question the "wisdom" of this reality would be unthinkable. The early American dominated car industry has been more or less usurped by the Japanese car makers who, in effect, have the world in their hands. Car manufacturing is spreading all over Asia; Korea, Malaysia and China have all entered the car business. Hyundai of Korea has turned out to be a rising star. There is no clear indication of when car makers will slow down their expansion. The expansionary trend is not only reflected in the number of cars on the road, but the production of more luxury models including big names like Acura (Honda), Infiniti (Nissan) and Lexus (Toyota).

The Corporate Good Deeds

It would be unfair to say that all corporations are paying no attention to the environment. The following are considered to be environmental-friendly products in areas of phosphate-free products: automatic dishwasher detergent, organic products such as meat, grain, fruit and vegetables produced without chemicals, biodegradable products such as garbage bags, energy-saving products like long-life and fluorescent screw-in light bulbs, chlorine free products (the brown paper) initiated by pulp-and-paper mills. Some of these products' contribution to environmental health is still under scrutiny by environmentalists, but their very existence is encouraging to the environmental cause. In addition, blue chip corporations including 3M, Alcan Aluminum, Imperial Oil, Northern Telecom and others are also doing their part to protect the environment. This shows that certain businesses, despite having profit-making as their prime objective, are investing in environmental care and this has not eroded their position in the market, nor discouraged investors from putting money in the enterprises. On the contrary, environmental-friendly products may generate additional profit for the companies, because consumers may express their support for the cause by buying their products. But all of these initiatives are still much the same as attempting to put out a forest fire using a garden hose. It would be most effective, if the accounting practice system is changed to include the cost of environment deterioration and the use of scarce resources. We could begin with the calculation of GNP and GDP.

6

Capitalism, socialism and entrepreneurism

Capitalism facilitates capital accumulation giving ownership to the rich. Communism claims everything belongs to the people, but gives ownership to the state. It is Entrepreneurism that satisfies people's inner need for creativity and innovativeness for the common good and gives ownership to the individual, an ideology propositioned to bridge the gap of social injustice created by micro economic theory between the greed of the individual and common good of the societal need.

"Ism"

"Ism" – three little letters that mean a lot to human history. The meaning of "ism" is no mystery. It is normally used in conjunction with another word, depending on the meaning of the other word. It could be very significant, describing a collective feeling about an ideology such as liberalism, spiritualism as well as those feelings in religions such as Catholicism. On the other hand, it could simply be a description of a disposition or human action. Capitalism, despite its glamour and attention given by the faithful, is not an ideology but a description of human behaviour and action associated with desire for the satisfaction of greater "want" for money or financial wealth, that is, the rapid accumulation of capital. Although capitalism has no proud origin or roots of wisdom, it nevertheless has prompted a historical nightmare experienced by people in every part of the world – the not so noble human experiences of colonialism and imperialism. (See Table 6.1.)

Table 6.1 An illustration of "ism"

Actor	+ "ism"	Remark
Imperialist	Imperialism	Political and military expansion. There is no element of common good for those conquered by the force.
Colonist	Colonialism	Capitalistic expansion through colonization. Perhaps it was good for the colonists, but not necessary for those who were colonized. In comparison with Imperialism, both were aggressions.
Capitalist	Capitalism	The process of capital accumulation for the good of the individual. Originated from biological desire for expansion, with little to do with the common good.
Socialist	Socialism	An ideology designed for the common good, with a distributive system-responsibility to the state. Unworkable in the human environment under today's circumstances, although some socialists' policies are implemented on a national scale by some governments.

The Development of Capitalism in Retrospect

From a scientific point of view, the origin of capitalism can be traced to a biological desire for expansion to maximize our habitat within the environment. Historically, capitalism was directly or indirectly associated with colonialism; the Dutch, British, French, Spanish, and Japanese colonization records are clear evidence of capitalistic expansion.

As capitalism has a long and direct association with human endeavours, over a period of time, a number of varied definitions were offered by concerned individuals; a few of the more noted examples are included in Table 6.2.

Table 6.2 Definitions or interpretations of capitalism in retrospect

Author	Definition	Source
E. Heckscher	Modern capitalism as that "unwholesome Irish stew".	*Economic History Review,* Vol. 7, p. 45
J. H. R. Carnwell and H. E. Czerwonsky	True capitalism means an economy of free and fair competition for profit and continuous work opportunity for all.	*In Defining Capitalism*
R. H. Tawney	As being no more than a political catchword.	*Religion and the Rise of Capitalism* (1937 edition)
Werner Sombart	The origin of capitalism is in the development of states of mind and human behaviour conducive to the existence of those economic forms and relationships which are characteristic of the modern world.	*Der Mederze Kapitalism* (1928 edition)
Max Weber	Where the industrial provision for the needs of a human group is carried out by the method of enterprise, and a rational capitalistic establishment as one with capital accounting; seeks profit rationally and systematically.	*General Economic Theory,* p. 275
Earl Hamilton	The system in which wealth other than land is used for the definite purpose of seeking an income.	*Economica,* November 1929, p. 339

Table 6.2 (cont'd)

Author	Definition	Source
F. L. Nussbaum	A system of exchange economy in which the orienting principle of economic activity is unrestricted profit. Such a system is marked by a differentiation of the population into owner and propertyless workers.	*History of Economic Institute of Europe* (1956), p. 61
Karl Marx	A system of production for a distinct market. The mode of production is not merely to the state of technique, but also to the social relations between men which resulted from their connections with the process of production.	*Capital*, Vol. 3 (1906), p. 914
G. Schmoller	The relation which exists between the production and consumption of goods, or the length of the route which the goods traverse in passing from production to consumer.	*Principles d'Economie Politique, passim (Mercantile System,* pp. 8, 9)
E. Lipson	The fundamental feature of capitalism is the wage system under which the worker has no right of ownership in the wares which he manufactures: he sells not the fruits of his labour but the labour itself.	*Economic History,* 3rd edition, Vol. 11, p. xxxvi
Evan Cunningham	A phase when the possession of capital and the habit of pushing trade have become dominant in all the institutions of society.	*The Progress of Capitalism in England,* pp. 24, 73

Table 6.2 (cont'd)

Author	Definition	Source
B. S. Kierstead	Capitalism may be considered as "the process of net capital formation". However, it requires that productive resources be re-allocated to 'tool-making' ... People with the monetary resources to buy the services of productive agents do so, and they direct these services into the production of capital. The people who accumulate the money resources and the people who make the decisions about what kinds of new tools to produce and in what quantities are not the same people. The first we shall call capitalists, and the second group entrepreneurs.	*Capital, Interest and Profit* (1959), pp. 5, 6
Raymond W. Y. Kao	Capitalism is legalized stealing from the environment and from the future resources of our children and children's children.	Based on Exxon's TV commercial claiming "We develop resources of the future for the present" or other commercials such as "We bring the future resources for your present enjoyment".

Capitalism is a Political Catchword

Among all of these definitions (Table 6.2), it is perhaps R. H. Tawney (1937) who has made the meaning of capitalism so transparent, since it is his definition that describes the present state. The former USSR, after approximately seventy years of playing with the slogan "working class revolution for the working class", has done its part to discredit Karl Marx's idea of communism and the socialist ideology. Keep in mind, though, that while communism is now a member of the "walking dead", it is not yet dead. While there are no suitable terms to describe the good spirit of "wealth distribution" and money stimulation for creation and productivity, capitalism seems to be a convenient and impressive enough catchword. Just look at the Group of Seven (G-7): an international superstar club consisting of seven powerful capitalistic nations, with no element of "communism" in its political process (though there may be some socialist schemes such as found in Canada). And to nobody's surprise, all club members practise a certain or similar brand of Western "democracy", though a few of them have queens or kings, who may say: "People in our country just love to have a 'Crown' on their 'head'." Therefore, "capitalism" really is the catchword, at least as spoken by political leaders. Friends in the academic community certainly do not wish to miss this "window" of opportunity, and must be credited for their "innovative" thinking by linking "capitalism" and "entrepreneurship" to make them two of a kind. As they say: "Capitalism stimulates entrepreneurship, and entrepreneurship promotes economic growth." It sounds very much like a slogan used in a North American protesters' parade, impressive and inviting. Therefore, as the "world turns", so do its "ideologies". "Capitalism" is the order of the day, year, century, and who knows, maybe forever. What a catchy phrase! However, if the question is raised in respect to the future of humanity, what happens if we have no other options and capitalism continues to be the "ism" that runs our lives? To predict the future is not a simple matter, yet Hollywood movies at times can provide surprising insight into the issues. For example, one of the movies (unfortunately, I am unable to recall its name) describes how a large corporation, with its financial power, acquired virtually every inch of the land of a city and owned all available resources. Everything within and around the city was all for the benefit of the corporation and its equity holders, with very little or no regard for the well-being of the people living in and working for the corporation. Someone in the city questioned the corporation's executives: "Why have you no concern for people? Your corporation is already so rich and owns everything there is to be owned, but still wants more!" The CEO of the corporation responded without hesitation: "We are in the market economy, you can always buy us off and be rich yourself." The CEO is right, but who has the power to challenge the ownership, since the people own nothing but themselves?

It may come as a surprise, but despite the commotion and fanfare, capitalism itself has no noble origin other than its direct and past association with colonialism. However, some thoughts have been given to what is supposed to be "good" about "capitalism". In essence, at least in the minds of faithful followers, capitalism is the free application of matter, energy and property (property is also energy in a different form) making things for trade in the open market, in an open society, where the ends are determined by individuals or by voluntary associations of individuals. In an academic context, it is a process of economic reality. The capital accumulation process, thus, may be illustrated as shown in Figure 6.1. The sprawl and positive feedback process of capital accumulation and desire for more can continue until it reaches breaking point.

Entrepreneurs are all for Wealth-creation and Value-adding

Since capitalism is prompted by human needs and wants, and functions in the human environment, the process of capital accumulation has never been meant for the common good. (The process of accumulation may help to create jobs initially: those retrenched by the corporations due to recession can always be self-employed by starting up their businesses, but the rapid accumulation of capital and high return on investment create pressure for the substitution

Figure 6.1   Development of capitalism

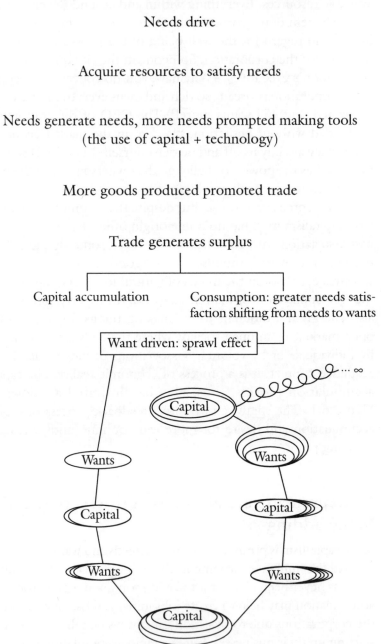

Needs drive

Acquire resources to satisfy needs

Needs generate needs, more needs prompted making tools
(the use of capital + technology)

More goods produced promoted trade

Trade generates surplus

Capital accumulation   Consumption: greater needs satis-
                          faction shifting from needs to wants

Want driven: sprawl effect

Capital ... ∞

Wants

Capital

Wants

Wants

Capital

Wants

Capital

of labour for capital, subsequently shifting to cost of jobs lost to society.) It is a self-governing and self-directed personal interest and human desire for greater "wants" that has sustained the capitalistic expansion. Capital accumulation itself has no element of innovation and creativity; hence, it has nothing to do with entrepreneurship. However, the availability of capital and the desire for a high rate of capital accumulation stimulate and introduce productive activities. In this connection, the process of capital accumulation and the process of wealth-creation and value-adding are two parts of the economy. One part originates by virtue of the desire for more money, and since money must be used to make more money that induces the second part of the economy: the entrepreneurial process. It was so noted by Kierstead (see Table 6.2). Those who hold financial resources for investment with the desire for making profit are capitalists and those who use capital to engage in the wealth-creation and value-adding process are entrepreneurs.

Socialism and Entrepreneurship

Socialism (including communism) as an ideology has indisputable value to humanity. Who does not want a society with no distributive problems? Everyone in this world would enjoy the fruits of his or her labour, and let those who are committed to work for the common good plan the production of goods and services to meet people's needs and the distribution of the harvest.

There are several parts to this socialist ideology. Some ideologies with the notion of a "just" society are inherent in the old (some people would consider it to be primitive) society. For example, in traditional Inuit society (natives of Canada mostly residing in the northern part of the country) people relied on fishing and hunting for a living. When old members were unable to hunt, they had their share of the action from the younger hunters' harvest simply by placing their hand on the animal, symbolizing their claim to the catch. It was a just system that worked well in a simple society. It worked because everyone respected the system, and the elders were hunters themselves when they were young. It should also be emphasized that the system worked well in the Inuit community, because both production and distribution systems were simple enough and a recognized tradition. Moreover, the general populace was more inclined to work to satisfy its "needs", and to a lesser extent "wants".

There is another incident that can be used to illustrate the matter of distribution. It is a simple incident, but very much explains why distribution is not necessarily a problem, if there is nothing or very little that can be distributed. It occurred in 1990, when I was giving a public lecture to a mixed audience. During the question and answer session, a participant asked me: "In your opinion, is there too much government intervention in the marketplace and too rigid regulation of business practice in our country?" I did not answer the question directly, but rather gave the following illustration.

Assume that this audience consists of approximately 300 people, with a mixture of children, and young/old males/females. None of us has had any food for days, so we are all hungry and desperately in need of food. Let us also assume that there is only one bowl of rice ready for consumption. What should we do with the rice? Should we divide the rice into 300 portions, each getting a few grains? Should we allow the young children to have the rice first? The women? The elderly? The one who has a gold and diamond Rolex watch? The weak? Or should we place the rice in the middle of the room and everyone would make a dash for it, so the strongest would have it all and the weak ones would have none?

The illustration was simple enough to comprehend. There is no satisfactory solution because the problem is not dealing with the matter of distribution, since there was so little or nothing to distribute. The real challenge then is how to produce more rice.

In a country which has little or no natural resources, government intervention in the marketplace is almost inevitable. The question is not whether the government should intervene in the marketplace, but how to intervene by putting the responsibility in the hands of the people, to elect representatives that know how to intervene in the marketplace for the common good. This is by no means condoning the practice of communism, because there is nothing to be "communed" about it. On the other hand, if production is the challenge of our society, then productivity rests on the individual. If the practice cannot permit individuals to identify themselves with the harvest (ownership) of their production, then there is little incentive for individuals to create. This is perhaps the greatest flaw in the communist ideology which led to the downfall of "pure" (or not so pure) communism, but not necessarily "pure" socialist ideology.

The greatest wisdom in the communist and socialist ideology is that there should be more equitable wealth distribution for the common good. Therefore, central planning in both production and distribution is the core. Whereas capitalism roots itself on the desire of the individual to freely apply resources for "private property rights", it sharply divides "rich" and "poor". Therefore, if the fate of communism is a nation's economy controlled by its government, capitalism, as we understand it and the way we practise it now, would be the reason for government intervention, unless we do not want to have a government at all.

In summary, while capitalism facilitates capital accumulation, giving ownership to the rich, communism claims everything belongs to the people, but gives ownership to the state. What is there left for "people"?

So far, we cannot find any element of common good in practising capitalism, and its flaws were discussed earlier in Chapter 5. Throughout this book, however, there has been one "plus" that would make capitalists and faithful capitalism supporters jump with both feet for joy – capitalism creates jobs and new wealth by inducing entrepreneurial activities. Even though it is not the intention of capitalists, the "job creation" reality, however, can only happen before the accumulation of capital reaches a stage of saturation. Once capitalists find they are unable to acquire a desirable ROI by creating jobs, they will begin to kill jobs instead. Socialism, on the other hand, is dedicated to the "common good", but fails miserably to appreciate the motivational factors associated with the direct linkage with the entrepreneurial initiative. By nature of the communist ideology, entrepreneurship, if narrowly viewed as merely being the "profit-making scheme that breeds capital accumulation", has no place in entrepreneurial activities, if "communism" is upheld as the centrepiece in the altar of socialist society.

Entrepreneurism: The Third Option

Entrepreneurism may not be easily found in the dictionary; it nevertheless has appeared in several published works. Most of these published works are focusing on new venture start-up, development or financing. Under the circumstances, these publications may be works of a practical value, but hardly make themselves an "ism", since "ism" refers to ideology. It is the ideological framework that encompasses elements of philosophy, an idea that can be implemented in the human environment.

If we place it in its proper perspective, entrepreneurism is an ideology based on the individual's need to create and/or innovate, and transform creativity and innovative desire into wealth creation and value-adding undertakings for the individual's benefit and common good.

Under entrepreneurism, an individual is a creative and innovative agent who desires ownership and the right to make proprietary decisions. A corporation is not a money-making machine, but a community of entrepreneurs created for the purpose of creating wealth for the individual and adding value to society; the state is therefore an agent to facilitate and make the environment feasible for the individual to create and/or innovate.

1. *The state: entrepreneurial government.* Under entrepreneurism, the state is the infrastructure consisting of individuals committed to serving people for the common good that will facilitate individuals to realize their right to economic freedom, their right to acquire ownership for the harvest of their labour, and their right and obligation to the protection of the natural environment.

2. *The individual: entrepreneurial person.* An individual is the centre of the economy, and as a stakeholder in any undertaking is responsible to himself or herself. This individual views entrepreneurship and working as an entrepreneur as a way of life.

3. *The corporate entity: entrepreneurial corporation and corporate entrepreneurial managers.* The corporate entity, regardless of its size, is a community of entrepreneurs. Individuals in the entity are entrepreneurs in their own right. They are all agents for change working together as a body dedicated to the creation of wealth for every individual within the entity while adding value to society.

Since the entity is a stakeholder in the community, its responsibility must not be limited to satisfying the shareholders' ROI only. It must, instead, satisfy all those involved in the operations of the entity, particularly the silent partner in business – the natural environment. Corporate managers under the circumstances are entrepreneurial managers in the corporate entity. In the words of Motorola University's advertising to attract students: "They are agents for change; talented, supportive and motivated." They are motivated because they are not merely inspired by the needs of consumers, but also individuals strongly motivated by a natural instinct for the entire operational process. (See Tables 6.3 and 6.4.)

Table 6.3   Entrepreneurism: The relationship between the individual, the state and the corporation

Corporate entity	Individual	Public: The state
A community of entrepreneurs committed to the common good to create wealth for individuals in the entity and add value to society	Responsible to him or herself. A stakeholder under any circumstances. Acquiring ownership through the harvest of the fruits of his or her labour	Committed to serving people, to facilitating individuals to acquire ownership through the harvest of the fruits of their labour and to the protection of natural environment

Table 6.4 Capitalism, entrepreneurism and socialism as reflected in the firm

The firm	Capitalism	Entrepreneurism	Socialism
Ownership	Shareholders	Individual	State
Individuals in the entity	Employees	Stakeholders	Employees of the state
Management	Representing the shareholders	Each individual is the manager of his or her own task	Representing the state
Sources of capital	Invested by investors	Investment plus sweat capital of individuals in the firm and/or through their share participating scheme	Supplied by the state
Competition for funds	Decisions made by investors on the basis of profitability	Investor's decision based on the firm's performance in the marketplace plus individuals' commitment to the firm's long-term growth potential	Allocated by people who have positions of power based on the perceived state priorities
Individual's remuneration	Based on the contractual agreement made between the firm (the management) and employee (individually or collectively)	A conciliatory process, with the possibility of workers' participation in the management, collective bargaining may not be necessary.	Determined by the state
Profit	Goes to the shareholders, bonus to top executives	Residual (on a better cost base) goes to shareholders	Belongs to the state
Profit computation	Revenue after all costs deducted (accounting method)	Residual concept, taking into account opportunity cost and cost to the environment	Similar to capitalism
Union's position	Bargaining with the management	A partner of firm's growth	Cooperate with the state

We must realize that both capitalism and socialism are merely processes in our economy; what matters most is to explore the possible options that will provide economic freedom and ownership that all individuals need. Whilst capitalism facilitates capital accumulation giving ownership to the rich, socialism (including communism) claims everything belongs to the people, but gives ownership to the state. It is entrepreneurism that satisfies people's inner need for creativity and innovativeness for the common good and gives ownership to the individual. Therefore, in our societal structure, entrepreneurism is the alternative to the two extreme ideologies, capitalism on the one hand and socialism on the other.

In conclusion, the following two arguments are necessary for us to further appreciate some of the facts related to the conceptual issues which are not so apparent.

First, it is claimed by capitalism faithful that today's capitalists are quite different from capitalists of the old days. In the 19th and early 20th centuries, a capitalist was both an industrialist and investor but today, as the financial market evolves to the advanced stage, it is possible for investors from all walks of life to invest in corporations. Now pension funds rank high in the investors' list, and behind pension funds are the retired workers. Therefore, capitalism faithfuls argue that in today's capitalistic society, corporate managers are sandwiched between two labouring classes. By confronting the powerful unions, they work for the workers and when the workers retire, they work for the "pensioner capitalists". As viewed by capitalists' advocates, this process is much the same as communism, except that under communism, everything is for the "working class", but in fact, workers had nothing of their own. In capitalism, the corporate profit goes to the capitalists who are workers themselves. Therefore, according to these advocates, why mustn't we promote capitalism, and let the market economy through invisible hands (pricing system) take care of itself?

However, the issue on hand is not who invest and who are investors. The focus is on how capitalistic expansion through corporate managers' activities in the marketplace affect the well beings of people, resources and environmental health that prompts the consideration of options needed to guide our obvious economic development.

Figure 6.2 Corporate managers sandwiched between workers of two different classes

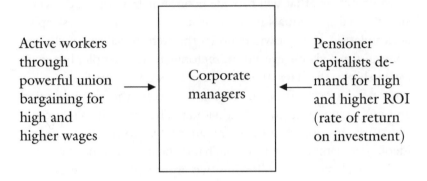

Active workers through powerful union bargaining for high and higher wages → Corporate managers ← Pensioner capitalists demand for high and higher ROI (rate of return on investment)

Then, there is the conceptual misunderstanding and/or error made by those against capitalism as it exploits the working class for the rich. Unfortunately, they fail to recognize the fundamental of capitalism that embraces three compounds, first, "capital", then "capitalist" and last, the "entrepreneur". We all have the clear understanding that capital is a factor of production, the sources to support and fuel our economic growth. Whereas a capitalist is an individual whose sole purpose is to use "capital", mostly to invest in corporate entities to accumulate capital through corporate earnings for personal financial gains. It is only the other group of individuals, hopefully all entrepreneurs (on the basis of the author's definition of who is an entrepreneur, Kao, 1995) through corporate creative and/or innovative activities create wealth (including a reasonable return on entrepreneur's own investment) and add value to society. Individuals in a corporation with the purposes of merely serving the capitalist's interest to accumulate capital with no value added to society are no entrepreneurs but employees of soulless and mindless corporations. To condemn capitalism by lumping the capitalist and entrepreneur together is the fundamental mistake of followers of Karl Marx including Marx himself. If Marx had made a distinction between "capitalist" and "entrepreneur", we may have a different world, and countless loss of lives and destruction of property could have been avoided. Marx may be excused due to the absence of well developed financial market at the time of his writing. As noted earlier, at that time, an industrialist was a man wearing two hats; a capitalist, on one hand, to create and on the other, develop the enterprise. In the traditional view, a corporation is a money-making machine created in the interest of capitalists, and it is the managers who, in fact, control the machine. Capitalists, in turn, control corporate managers through the mighty three letters: ROI. Although this is how it is perceived, it may not be so.

7

Unifying theory and corporate vision

The value of a 50-year-old tree, based on an assessment by Professor T. M. Das of the University of Calcutta, is:

Provide oxygen	US$31,250
Air pollution control	62,000
Soil erosion control and soil fertilization	32,250
Provide water	37,500
Animal recoveries	31,250
Total value	US$194,250

(Source: Singapore Zoological Gardens. Origin: Update Frosty, Michigan State University.)

The above does not include value generated through the use of timber, fruit for consumption, medicine for health and aesthetic value.

Corporate managers and accountants, have you ever accounted for or thought of including the above into the cost of your operation or in providing professional service before deriving your "profit"? How many trees valued at $194,250 each should be added to the cost before you decide on corporate profit? If you did, would you still be chopping down countless trees in the Brazilian forests for raising cattle to make hamburgers, or build golf courses in the countryside?

Memories are Empiricals on the Wall, which is the Closest to the Reality of All

The author has no religious affiliations, but has always attempted to live a life of love, not just for individuals (although that is where it all starts) but for humanity and the world as a whole. The author's passion for life has been nurtured and developed through hardship and good fortune over the course of a lifetime.

His childhood saw some of the hardest times humanity has ever known. Fatherless at the age of five, he soon found himself in the middle of the Japanese invasion of China, and spent much of his pre-teen years as a refugee. He witnessed the Nanjing massacre of 1937, and numerous atrocities committed by the Japanese imperial forces – whether or not these actions matched or exceeded the horrors of the Holocaust, no one can say, but certainly these events never received the attention and public sympathies of their more infamous counterparts in Germany. Poverty, hunger, sickness and an embittered refugee life, compounded by witnessing military struggles between the old Nationalists and the Communists[1] and muddled by different ideologies, prompted him to question the meaning of life and the purpose of being alive or living.

Yet all these problems also had some benefits. In these hard times the author often found it necessary to look for solutions to problems in new and unorthodox fashions, answers found outside the structures of formal learning. As a refugee in Chungking (southwest China), he was exposed to a variety of ideas, histories, philosophies and ideologies in the local Xing-Hua bookstore and YMCA libraries. He was not only driven by a desire for knowledge, but also by more practical means – the bookstore and libraries were the only places in the city that a homeless person could spend long hours without being chased away by the proprietor.

It was this learning from the original sources, rather than knowledge disseminated through the intermediaries of formal institutions, and of teachers taught by teachers taught by teachers, that the author became convinced that it is profit that makes the world go round, and it is the favourable result of trade (business) that generates profit. "Profit" is in some ways a beautiful word, the driving force for economic growth and the source of employment.

On the other hand, the unfortunate side effect of trade is human misery, at least, when it is applied without thought or restriction. The desire for rapid accumulation of wealth and attainment of financial objectives has often driven business to an extreme, often an extreme of war and violence. Though people fight wars, it is the corporations and capitalists that start them, and it would be impossible to deny that without the backing of industry, neither Germany nor Japan would have started the Second World War. Cows are blamed when rainforests are destroyed to make way for pastures to feed them. But as the painter Isidro Con Wong of Costa Rica shows in his painting *The Dance of the Innocents*, the cows are not to blame but the logging companies. It is not only the logging companies to be condemned but mostly those wanting mass marketing and selling of hamburgers in the name of "profit", causing the deterioration of nature. In the same way, soldiers going to war are not to be blamed for the war – it is the ones who send them, and the ones who profit from their going.

Moreover, there is another issue that few people seldom trouble themselves to explore: if there is a "profit", where was it derived? Of course, we can always find the meaning of this word from the dictionary, but this is not the point.

Following the Second World War and before the communist takeover of China, the author migrated to Canada. There, over a period of two years in a now formal learning environment and in one year of a close working relationship in research with the late Professor Kierstead, the author got deeply interested in searching for an answer to the issue of "profit" – what makes people think there is a profit? We know perfectly well, from the point of view of our earth, we cannot add anything to the resources around us. We can only use and modify them. True enough, with different combinations, we increase the utility functions in the resources we acquire, but we have not been able to prove to ourselves that we have created more than what the planet initially provided for us. Then, where is our profit? If we adopt Nobel laureate J. R. Hicks' definition of profit, humanity as a whole should be as well-off at the end as we had been at the beginning, but are we? Nevertheless, there might be perceived "profit" for the individual corporation or an individual and his profit. If closely examined, it could only be derived through the following:

- Manipulation, as we have learned from accounting, arbitrarily determining what cost is. Only recognizing the cost what accounting perceives to be, but not what real cost is.
- Taking what belongs to others; the general working populace are paid less than their labour's worth (this is perhaps why some people believe in communism).
- Taking from the future; since our future generations are yet to be born, they cannot protest.
- Taking from our environment; it is our silent partner, but represented by silent protest through deterioration, death and extinction.

This observation convinced the author that the function of any economic undertaking is the allocation of resources so that humanity as a whole can live and prosper together, with every individual given the opportunity to make decisions to direct resources for the purpose of creating wealth and adding value for both the individual and society. Only through this form of economic process can we have a society of entrepreneurs who are all committed to create and innovate. Our businesses and corporations, in particular, have done nothing more than re-allocate resources, make different combinations, and to the best of their ability, increase the utility function of various resources. This leads the author to question the whole purpose of a corporation. Do corporations make profit to satisfy their executives and investors or to serve people as they often claim in the common motto: "To give our customers the highest satisfaction"?

As living creatures of earth, humanity, in reality, is living on borrowed time. We could survive and prosper as a whole for a few million or even billion years, or maybe just a few hundred years, depending on how we manage our resources. Some of us believe that what is on earth is there for us to use and to take from the future for our current consumption. Yet at the same time, most believe that humanity must continue. That is why we have and love our children. But how can we survive, and have children as we do, if we take away the life-support system that belongs to them? We do have a stewardship function on earth to safeguard resources for future generations.

The question narrows down to the matter of resources, wealth and value, not profit. Our corporate managers must come to the realization that it is the intelligent allocation and management of resources which is the only justification for their existence. Wealth, for all intents and purposes, is an extension of the individual's decision-making capacity through the acquisition of ownership over resources and intangibles such as knowledge, wisdom and improved human relations. Value is a utility function which is the result of individual or collective innovation and creation activity adding to the common good. On the other hand, "profit" as a measure of corporate performance must be an indicator that reflects the management of resources in the interest of the common good, and is a matter of creating wealth and adding value for both the individual and society.

In Retrospect

$F = Ma$

Some five years ago, a friend of the author complained about an unusual action taken by his parents. According to him, one day, he got a call from his mother telling him that they (his parents) had decided to get a divorce. His father was 83 and his mother 77. The call shocked him and his family. He immediately telephoned back his mother asking her why they had decided on a divorce; the mother refused to say anything more than: "It's final." He was desperate to see his parents, if for nothing else, to find out what had happened to his parents at their age to want to divorce after a marriage of over fifty years. Upon arriving at his parents' home, he found a party (very much like a New Year's Eve party) in progress. The living room was filled with people; brothers, sisters and their families. There must have been at least twenty people talking and joking and having a good time. He asked his mother: "What's going on?" His mother laughed and said to him: "We are not getting a divorce. We thought this is the only way we could get all of you to come." The truth was, his parents at their advanced age wanted to see their children and grandchildren, but everyone was so busy with their own priorities. Once in a while they all did give the old couple a call, but even the phone calls were getting less frequent because it was felt they were too expensive. The parents used every trick in the book to get their children to visit; they told them their father was sick, or their mother had injured her knee. All the parents got was simple advice such as, go and see a doctor; or take a lot of aspirins and have a good sleep. Finally, they resorted to "divorce". The news shocked their children and brought them home. Just what the parents wanted!

This little story serves as a reminder of one of Newton's fundamental laws of nature (gravity and motion): every object at rest remains at rest until acted upon by an external force. It can be expressed as:

$F = Ma$
where F = force, M = mass, and a = acceleration.

These parents certainly knew how to use this fundamental law to get what they wanted; employing shock tactics to push their children out of their routines and get them to visit. The challenge for the parents is, what's next? There are two possibilities: one, their children will realize that parents need to know that they care, and visit them often; the other is, they will become wary of falling victims to such parental games again, and make sure there won't be a next time. Whatever the case, it was certainly more effective than doing nothing.

In a similar vein, the first six chapters of this book have been designed to be provocative, like the old couple's "divorce" story, and in accordance with Newton's law. Without an external force, business management and business managers would still and always will be at rest.

This is perhaps the reason management discipline has seen so few new discoveries. After all, if life has been "good", why trouble to make any fundamental changes? Worse still, most of the time any signs of a new thought are immediately quashed by the big "guns" (or at least those who think they are) condemned with cries of "too descriptive", "no empirical evidence", "how do we expect corporations to make a profit", and "without profit what's our purpose for being in business?" And of course, there is always the battle cry of the apathetic: "don't rock the boat".

And yet the empirical test, the holy grail of business research, is often nothing more than limited data collected to reflect a small segment of the happenings of a large community, and usually more a reflection of the past than a projection to the future. What about Professor Das' valuation of a 50-year-old tree? Empirical or descriptive? Can anyone put a value on his or her work or should we just simply dismiss it as the notions of a "madman", and nothing more than unsubstantiated rubbish?

A short return to chapters one to six

The first six chapters were devoted to the following objectives:

1. To realize that a corporation is not a dream money-making machine, but a community of entrepreneurs (an entrepreneur as defined by the author).

2. To appreciate the fact that there is no such thing as "profit", unless we have restored the capital used for the process.

3. To challenge our "B" school education; the whole business education has no philosophical roots, but is based on a general management model that is essentially governed by the theory of the firm, the rate of return on investment and regulations issued by stock exchanges.

4. To recognize that both the environment and social issues are corporate responsibilities, not just the government's.

5. To take a closer look at the cost of capitalism, and how it impacts humanity, the environment and resources.

6. To recognize that both capitalism and socialism are economic processes, while it is entrepreneurship that stimulates the individuals' inner need to create and innovate, giving a purpose to life for individuals and promoting the common good.

Beginning with this chapter, all efforts will be directed to stimulating the corporate managers' thoughts, to making decisions and acting as entrepreneurs.

It should be noted that although no significant progress has been made in the development of the business management discipline, there are concerned individuals who have made various attempts to reshape some matters that affect corporate management practice. Table 7.1 is an overview of these efforts.

Table 7.1   Concepts and practices as perceived by the author since the Industrial Revolution

Discipline	Attempted changes	Changes suggested
Accounting financial reporting*	Measurement for profit based on agency theory; productivity measurement and added-value approach.	Recognition of stewardship function extended into limited resources measurement and accounting. To replace the use of "profit" for financial reporting by a more meaningful measurement such as: value added.
Accounting: Cost recognition and measurement *	Some environmental costs; pollution, water usage, social responsibility.	Real cost, opportunity cost, cost of present and future.
Economics	No changes meaningful, the inherent theory of the firm, continues.	GNP, GDP measurement, use resources management approach.
Industrial structure	From Industrial Revolution to factory, from factory system to branch plant operation to offshore subcontracting to regional based operations, home based SME's (small and medium enterprises).	

Table 7.1 (cont'd)

Discipline	Attempted changes	Changes suggested
Management	No significant changes, use of bottomline approach to measure managerial performance, and the use of staff control and oversee the implementation of rules and regulations, the use of consolidation of financial reporting.	To have a global vision (present and future) and use an entrepreneurial approach to manage but not to dominate our limited resources to achieve wealth and the value-adding objective.
Ideology: Capitalism	Evolving from its association with colonialism, imperialism to capital accumulation of the private sector of economy.	Capital as a utility function, for allocation of limited resources to create wealth and add value for both the individual and society.

* See Pang (1995, pp. 25, 102–3, 112, 131, 233, 245–90), with special reference to Williams (1995).

"To Grow a Branch from 'Roots'" Wisdom

A branch grown from the roots is stronger than a branch grown from another branch.

It has been a well-established fact that in this day and age, academic institutions have established a simple unwritten rule that governs all academic research work, including academic research papers and dissertations of higher degrees. If a paper or dissertation is expected to pass through the referee's review process or external examination, the researcher or candidate for the higher degree must observe this simple rule. There are two parts of the rule which cannot be found in writing anywhere:

1. It must have references, the more the better; a paper with five pages of references is better than a paper that has only five references. The same goes for dissertations.

2. The reference list must include a citation of the works of the referee or external examiner. It is even better if the work is quoted verbally during the examination.

The above observation was discussed with two professors (let us call them David and John) from a well-known European university, at the VIth ENDEC Conference organized by the Entrepreneurship Development Centre of Nanyang Technological University, Singapore, and held jointly with Shanghai Fudan University. The issue was briefly discussed but with no conclusion. David remarked: "If this is how we develop our best people in the management discipline, the only possible thing that could happen is to turn out better copiers, refining what has been said and written." John agreed with David, and added: "It seems what we want our researchers and Ph.D. candidates to do is to read and cite some recent works, and not really to encourage them to research into original works of great scholars." The whole exercise is very much like growing a branch from a branch, then grow more, newer branches from the newly grown branch. Consequently, the branches get thinner and weaker, very much like fruit trees, with branches and leaves growing from the branches, but bearing no fruit. This is what is happening in the business management discipline. Perhaps this is why there are so few breakthroughs, no new thoughts or knowledge. All we can expect is, perhaps, different versions of the same story, much like a broken record, playing a few pieces of music, over and over. David furthered his concerns about teaching in business management, noting:

> What concerns me the most is that we don't encourage our students to think, but reward them to repeat what's in the book, what has been taken down from lectures. We examine and test them, and the ones who get high marks are the ones giving back what we have taught, while those expressing new thoughts or ideas, more often than not, are penalised. When examiners mark papers they based their evaluations on standard solutions. Let's say that he or she starts to mark down the answers considered to be unacceptable, and then deducts the wrongs from the base of 100; and it wouldn't be surprising if the marking scheme leads the examiner to give the student a negative result, you owe me twenty-seven marks, so to speak!

He finished his remark with a laugh.

The conversation in Shanghai with David and John reminds the author of a personal experience. About thirty years ago, he was the only student in "Profit Theory" supervised by the late Professor Kierstead. Professor Kierstead was seriously ill in hospital, where he conducted a "one-to-one" teaching session. Professor Kierstead was an established author in economics even at a young age, but the reading list contained absolutely no works of his own. The author asked him why. He said:

> Raymond, just keep this in mind, a branch grown from the roots will be a strong branch. If your purpose is the pursuit of knowledge, do research into original works, not mine. Mine is not original knowledge. I don't want you to be another Kierstead. Cultivate your own thoughts, read the original works and develop from there. By all means read other people's recent works, but bear in mind, some of the works are built on top of other people's work, and possibly, on top of someone else's work again. It is your own thoughts that will contribute to the body of knowledge, not the works of others and on top of others', including mine.

When the author recalls the experience, how he wishes Professor Kierstead were still here.

Corporate Vision

"Vision" used for business management is a description of the corporate executives' perception of the corporation's position in the marketplace. Although there are people who feel that the organization's vision differs, or at least should differ, from a personal vision, its development and formalization into a comprehensive (or simple) statement are very much the work of individuals, which cannot escape personal beliefs or disbeliefs. It is the work of a "human" telling other "humans" the corporate prediction of what the future holds.

Corporate vision "TQM" (Total Quality Management) style

There is a great deal of variety in the statements of vision among organizations; some commonly noted ones include the following (Goetsch and Davis, 1995, p. 109):

Customer-driven vision:

- We will be recognized by our customer as producer of the highest quality _____ in the world.
- We will be seen by our customers as the suppliers of choice of all our products.

There are also statements which are listed in the same service as unacceptable:

Profit-driven vision:

- Our reason for being is to make money for our stockholders.
- Our ultimate objective is to maximize profits.
- We still build a revolutionary new _____.

The former type is a clear indication that the corporation attempts to project a customer-driven image, and the latter is clearly a vision statement for "profit-driven" or "capital-driven" companies. If the former is for the public, the latter would be a suitable statement for board meetings behind closed doors. Why is it not possible to have one statement to serve two purposes? While this is a lovely dream, it will never happen so long as the principal concerns of the board members (representing the majority of shareholders) are their profit requirements; and as long as they can threaten to pull their money out of the company and invest it somewhere else. Therefore, it appears that the greatest challenge for our corporate executives and managers is to find a way to bring all these "drives" into a unified approach to develop a vision that sustains the corporate purpose while satisfying both customers and shareholders. (See Figure 7.1.) The simplest expression is:

If our customers recognize us as the highest quality producer, through buying our goods (or services), we will make the profit we desire, thus providing a satisfying return to our shareholders thereby inducing additional capital investment for our continuing growth and expansion ...

Amen. What a convenient approach!

Figure 7.1   The roundabout challenge

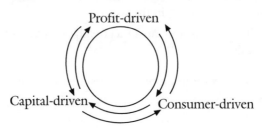

Perhaps there are some companies that take this enlightened approach; on the other hand, there are countless corporations that for reasons of satisfying corporate objectives, use advertising and promotion to tell consumers: "This is what you need, our product is the best in the market." The manufacturers of the drug Valium have long been aware of the potential damage to the users' physical and mental health, yet have been unwilling to pull out Valium from the market. Similarly, the car makers with full knowledge of all the faults of their product continue to promote the unrestricted sale of more and more cars. They have made only minor improvements in their environmental fitness under great pressure from the government and the people. The corporations who should have been the leaders, have only been the most reluctant of followers. Are these corporations in business for profit or for consumers? Perhaps it is for both, but which comes first, profit or consumers?

An illustration of macro–micro approach to corporate vision development

There have been numerous ways of developing a corporation vision. If it does not begin with imagination, but is based on the man-

Figure 7.2 A common macro–micro approach to corporate vision development

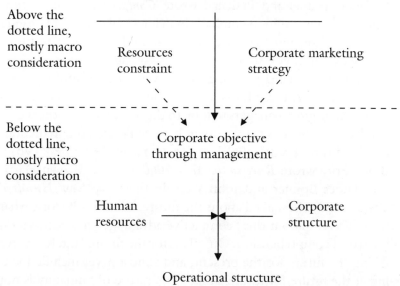

Corporate vision expressed through a vision statement

agers' need to attain the corporate objective, it may be expressed as shown in Figure 7.2.

- Corporate vision (man-made):

  The only environmental factors that are important are those that immediately affect the corporate need to attain its one-year to three-year profit goal, which seldom extend beyond five years. Under the circumstances corporate executives are likely to concentrate on personal benefit. For example, how can I move up to the 99th floor (executive suite) and get an executive parking spot by the year ...?

  The vision is normally based on the firm's research, publicly available information (from macro-scanning, and including such recent sources as the Internet), prior knowledge of consumer behaviour and predictions of the availability of needed resources in the near future.

- The reality of corporate objectives is the bottomline of a quantitative expression.

To secure a foothold in the future

In 1994, Hamel and Prahalad wrote *Competing for the Future*, which initiated a shift in emphasis in managerial education, urging corporate managers to secure a foothold in the future. This futuristic view involves such methodologies as macro-scanning, forecasting, scenario planning and developing technology foresight. However, these ideas are far from new. All along, writers have been urging managers to look to the future; in fact, the whole emphasis of business management has been placed on predictions for the future. Fred L. Polak wrote *Image of the Future* (1961), Baade Fritz wrote *Race to the Year 2000* in 1962. Brown Harrison, James Bonner and John Weir, in their *The Next Hundred Years* (1958) also talked about the future. With this history, what is so different about this "completive advantage" futuristic vision is beyond comprehension. Our plans for the future largely involve taking the future for the present, and almost never include looking at the future for future's sake. The future of humanity is not just about "technology", but about everything that affects us now and in the future; in particular, caring for people resources utilization, and concern for the environment and all the ecosystems that surround and support us.

The Use of a Unifying Theory for Corporate Vision Development

The unifying theory

We all are part of the planet Earth. Earth is energy itself, as we are ourselves. Human activities represent the flow of energy, and business management is, in fact, the management of energy flow. It follows that a corporate vision which is used to set a course of action is not complete, unless it has its roots in nature. Therefore, physics, ecology, biochemistry, chemistry, biology, the humanities and all other fields of study should be a part of the consideration in the process of corporate vision development.

A corporate vision with roots in nature

The solar system is only a minute fraction of the universe, yet it encompasses all our activities. In contrast corporate management is only a fraction of human activities in this environment. A full understanding of nature is beyond the scope of this book; indeed, it is beyond the author. Yet, a corporate vision cannot be meaningful if it is not developed within its roots in nature. It must be acknowledged that science and the humanities need to be part of corporate vision development. What businesses do affect our total environment, including ecological health and the use of resources. They provide the impetus for the search for new frontiers, both for knowledge's sake and the more practical development of new products and technologies; last, but not least, we must never forget their impacts on individual people.

A corporate vision with roots in nature may be illustrated as shown in Figure 7.3.

1. Science

(a) Physics – the origin of origins

For the purposes of learning and convenience of research, physics has been arbitrarily divided into a number of areas, and often interrelations between areas have been missed. For example, both electricity and magnetism are part of the electromagnetic spectrum, but we study and learn them separately. It was the discovery of their interrelationship which in recent decades has led to the development of micro-electronics, one of the great triumphs of physics, and thereby, natural science.

During the twentieth century, discoveries in physics have had dramatic effects on our lives. The discovery of atomic radiation, originating in the works of Marie and Pierre Curie, has led to massive changes in many fields, including medical technology. The development of quantum mechanics by a host of physicists, including Einstein, Dirac, Schrödinger and Heisenberg, has led to

Figure 7.3　A corporate vision with roots in nature

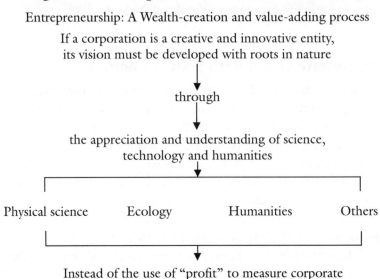

Entrepreneurship: A Wealth-creation and value-adding process

If a corporation is a creative and innovative entity,
its vision must be developed with roots in nature

through

the appreciation and understanding of science,
technology and humanities

Physical science　　Ecology　　　Humanities　　　Others

Instead of the use of "profit" to measure corporate
performance, the value-added approach will be used.

solid state physics and eventually silicon chip technology to practical application. The recent discovery of information and communication technology inherent from electronics is conducive to the application of the Internet.

Physics continues to work for our benefit, facing the challenge of finding new technologies and sources of energy to support industrial expansion and consumers' needs. Under the theory of "what destroys can also heal", the laser is becoming an increasingly popular tool in surgical operations. Magnetic resonance imaging has revolutionized non-invasive diagnostic techniques. Outside of medicine, the exploration of space, incredible advances in global communications and the continued development of new energy sources such as geothermal and solar are the result of physics research. If the idea of "system thinking" is to be used for corporate vision development, an understanding of physics and its resulting technologies, even at a basic level, would help corporate managers to know the great system of nature embedding this knowledge and would give corporate managers a broader understanding of what is around us; nature at work and how it affects us now and in the future. All corporate managers, without exception, must make decisions not just to please stockholders, but more importantly, to serve consumers and future consumers as well. The wisdom of "in harmony with nature" has everything to do with people's mindset, behaviour, and most importantly, decision-making for other people and for nature.

(b) Biochemistry and chemistry

While physics is the most fundamental of the sciences, it is the fields of chemistry and biochemistry which have had the greatest impact on industry, through the utilization of resources and the creation of consumable materials. We rely on fossil fuels for most of our energy needs. It is the knowledge of our dependence on these exhaustible resources that has led to the search for alternative energy sources. At the same time, we seek to understand the flow and transfer of energy within the cell, and both types of consideration should be a part of the vision necessary to sustain corporate purpose in society. Since they are all concerned about the allocation of resources, and all partake of a creative and innovative process, to develop a corporate vision it is no longer sufficient just to scan the economic movements around us, but to embrace nature that is around us as well. While scientists tell us about nature and how nature is at work, it is corporate managers that manage nature to increase its utility function for humanity now and for those yet to be born.

2. Ecology

Ecological impact should always be a consideration for corporate executives. To maintain the ecosystems that support the living is far from a casual matter, yet it is often viewed as little more than a topic for after-dinner talk. There is graphic expression of the ecological crisis at hand expressed in the author's book *Entrepreneurship: A Wealth-Creation and Value-Adding Process* (1995, p. 120): "If a lily pond contains a single leaf on the first day and every day the number of leaves doubles, when is the pond half full, if it is completely full in thirty days? The answer is twenty-nine days." Human beings are on our twenty-ninth day on earth, we have only one more day to ensure our life-support system can continue.

No single person could heal the damage we have done, because it is the greatest problem we have ever faced, a problem for all mankind. Awareness of this problem has permeated even our pop culture! Take for example Michael Jackson's video presentation Earth Song, showing dead bloody tuskless elephants lying on the ground in the jungle. Tree stumps cover a once vibrant forest floor, and other lifeless animals fill the screen; another scene consists of people crying, on their knees, hugging the soil, pleading for mercy, perhaps from "heaven", and asking the question: "What about us?" It was a moving presentation, but it is not heaven that has caused the problem. It is a problem rooted in every part of the world, caused by "profit-driven" corporations, and by individuals – those seeking ways to survive and those living in comfort and luxury, including Michael Jackson and the author himself; all are sources of this great problem. On the other hand, conserving and restoring the ecosystems must not be the work of one person, one group or even one nation's efforts. It is the work of all of us, and especially of the corporations, who until now have been followers, and must become leaders, in preserving and restoring the environment around us, as well as lessening our demand on nature. In any part of the world and under any system, we must account for everything: flowering plants and trees, all species of birds, animals, and even (perhaps especially) the microbes and bacteria – for they are a vital part of nature as well. We must care for frogs, snakes, lizards, insects, deer, bears, sea lions, whales, fish, etc. Everything and everyone must be our responsibility, particularly corporations' responsibility.

It goes without saying, there is much in life science to be learned; in particular, the behaviour of other living creatures can be a source of knowledge to help corporate managers to appreciate the behaviour of others either within the corporation and/or in the marketplace. For example, even without computers to tell them about "system thinking" and without the knowledge from quality management or "total quality management and beyond", ants and bees live in near perfect harmony – models of system thinking and cooperative behaviour. Killer whales feed on sea lions; they conserve their numbers carefully, always managing to preserve and maintain adequate food supplies for their young and the future.

Perhaps we all have to learn about preservation of our resources, because if we continue to expand our consumption rate pushed by corporate needs for expansion, what will the future be?

3. Global scanning: What's up the sleeve?

During the years following the Second World War, the Western nations (including the US, the UK, France and Germany), perhaps due to developing countries' inability to compete, filled the marketplace with high quality goods. The standard of living in these countries was by large much superior to the developing regions and countries, in particular, Japan, Korea, Taiwan, Hong Kong, Singapore and other Southeast Asian countries. By the 1960s, there was a surge of industrial activities in the less developed regions, initiated by the early blooming of the Japanese economy, assisted by the US, which made Japan the first small dragon out of the incubator raising its head to the world. Then, Korea, Taiwan and Singapore, modelled pretty well on the Japanese strategy "learn from the greatest, but be the greatest of the greatest" added to the little dragons team. Hong Kong benefited from China's Cultural Revolution. The revolution saw an exodus of a large number of "well-off" Chinese from large cities such as Shanghai and Guangzhou and resulted in a massive number of unofficial political refugees. Hong Kong, a not so new little dragon, wagged its tail and joined the little dragon league. Those little dragons suddenly caught the world's attention.

At the same time, in emulation of the West and most significantly the big G-7, the dragons were inspired to develop the same high standards of living and to covet the same comforts and luxuries. As we moved from the 1960s to the 1970s, China, the "mother dragon", marked the end of its Cultural Revolution (around 1976), and sent its "little red guards" back to school once again. Its new leader, Deng Xiaoping, outlined plans to set China back on the road and to a market economy. The idea of practising a market economy within the socialist system caught the West by surprise (at times in disbelief), and created an enormous window of opportunity: investing in China. Why not? If nothing else, China has the largest potential consumer market in the world. Labour is cheap, people are hard working, resourceful and above all, it has plenty of natural resources. So through the late 1970s until the Tiananmen Square incident in 1989, China looked like a gold-mine to investors worldwide. Capital flew in by the billions from Hong Kong, Taiwan, Singapore, Japan, the US, Germany and other countries (in contrast to Dr Sun Yat-sen's description of China as a sub-colony) like bees around a blossoming flower. This marked the completion of the dragon story. Now six little dragons play in the bluest sky and under the brightest sun around the mother dragon, awake but still yet to stretch its legs, lift its head and flap its tail. Instead of goods being produced in Japan and Western countries, competing and selling in developing countries as in the old days, now the tables have suddenly turned. Goods made in Korea, Taiwan and Hong Kong, China and later other countries (but carrying Western or Japanese brands) have flooded the world market. Anywhere you go, you will note goods sold with Asian origins, including local souvenirs made to please tour-ists: "Take a piece of Canada, the US or Britain" with you, but often made in Korea, Hong Kong or Taiwan.

This period of economic transition coincided with the Cold War period when the Warsaw Pact and the North Atlantic Treaty Organization were two powerful groups shouting at each other, much like two angry men posturing and threatening before the real fight begins. Perhaps it was a God-given miracle; even though the shouting matches were very sensational, the "fighting or kill-ing match" was never realized. Instead, the end of the Cold War accelerated a process which had already begun on the European side. A number of countries developed from the "Treaty of Rome", came around the idea of a European Common Market, starting with free trade, and evolving into a "European Union". (See Fig-ure 7.4.) What follows is the theoretical background in brief:

- Human rights are guaranteed by a European Charter.
- Conventional capitalism supported the Union concept.
- Nations form only an economic union and do not prevent indi-vidual nations from practising protectionism, but with member nations in competition for a better government.

- Inefficient or poorly thought-out regulations were eliminated, and property rights simplified.

- Member nations under the Union remain sovereign nations, but are member nations of the European government as well.

Under the Union concept, citizens of all member nations would hold European citizenship. There will be a common currency (an exchange rate will be determined in 1999), a European Parliament, a European Central Bank, and one political entity. As the German Chancellor Helmut Kohl said: "Germany is the fatherland, but the European Union is the future." (These may not be his exact words, the author accepts the responsibility for the paraphrase.) So, the present European Union, from an original handful of countries, has at the time of this writing, become fifteen member nations strong and is still growing. Although membership has not yet been extended to the Middle East, or even to include all European countries, the key players have made it very clear, as stated in its theoretical foundation, that others are welcome to join the Union; however, a would-be member nation would have to subscribe to the "Union" charter and any other Union rules. Of course, there are other step-by-step requirements.

Figure 7.4   Civilization on the march: From killing to trade, the trend of world politics

War ⟶ Cold War ⟶ Protectionism

↓

Regional trade agreement (differentiate tariffs for different countries)

↓

GATT (General Agreement on Tariffs and Trade)

↓

WTO (World Trade Organization)

The idea: The utmost goal of market economy – free flow of goods, services and people through agreeable prices.

In Southeast Asia, about thirty years ago, the idea of forming an Association of Southeast Asian Nations (ASEAN) was incubating. First, it was five countries: Singapore, Malaysia, Indonesia, Philippines and Thailand, then Brunei made it six. Now with membership up to ten countries, the association is finally maturing with a similar, but less extensive, mandate to the European Union (no provision for unified currency, but member nations are making an attempt to stabilize their currency value, citizenship, central bank, or parliament). Their policies include trade and environment protection, defence and nuclear weapon bans, which puts them one up on the European Union – France's nuclear tests in the South Pacific have caused who knows what damage to the environment. But the French politicians got their way. And for what? France has always felt the need to assert itself, not to bow to any nation; in some ways this is admirable, yet I think the cost of this form of glory is too high to the world, particularly to the environment. But what can anyone do? After all, France is a permanent member of the Security Council of the United Nations.

In North America, the US pushed through the Free Trade Agreement with Canada; the two then jointly got Mexico to form the North American Free Trade Agreement (NAFTA). These could easily become the early steps towards the completion of an American Free Trade Agreement, once nations from South and Central America join the league. But since the free trade agreement is only about trade, NAFTA is not anywhere near to the scope of the European Union, and is not even comparable to ASEAN. Even though the theory is obvious, the reality is still not so easy, because of the lack of economic stability of countries in Central and South America (even Mexico); it makes the union twice as difficult to create as the other two unions.

Despite these teething difficulties, NAFTA, together with the European Union and ASEAN, form the big three trading blocs which will determine our economic future well into the next century. Aside from these three big groups, there are also at least two dozen or so other trade groups formed for the purpose of dealing with trade challenges. And APEC (Asia-Pacific Economic Cooperation, including countries such as Japan, China, Australia and

New Zealand), which began with very unclear goals of broadening US involvement, has strengthened its position to become a significant player in the world market. Free trade or no free trade, the world is certainly better off dealing with economic issues than using nuclear warheads or pulling guns to shoot each other. It should also be noted that there are speculations of a possible joint development undertaking between India, China, Korea and possibly Japan and Russia. (See Figure 7.5.) If so, it would make the new group the largest economic cooperation group on earth on the basis of population and land area. It should also be noted that top brass from the European Union and ASEAN are seeking possible means of cooperation. With APEC and "Euro-ASEAN" (a term invented by the author) hand-in-hand, even though it may not live happily ever after in holy matrimony, these preliminary overtures and the body language alone would make people all

Figure 7.5   The perceived trade cooperation between China, India, Korea and Russia and possible inclusion of Japan and ASEAN

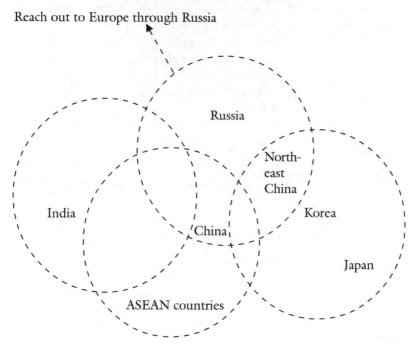

Reach out to Europe through Russia

over the world wait eagerly for the big celebration: Year 2000, a landmark of human history. That will be the day!

The world players have evolved from interacting through military to political then economic means. The bottomline, if there is one at all, is to make all our resources flow freely from one corner of the globe to another, thus making resources efficiently and effectively used in the interest of humanity. (See Figure 7.6.) The day when countries all over the world eliminate their immigration and custom departments from the governmental structure would be a day of celebration. Would it be possible? We do not know. We know that people are working on it, but in contrast, there are just

Figure 7.6   Scanning the world: United or sharply divided?

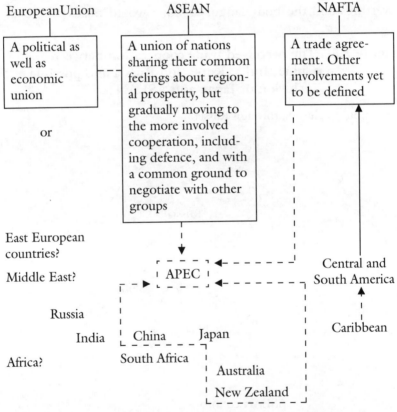

Note: Countries that are outside the dotted line or solid line block are not officially affiliated to any particular groups.

as many, if not more people, whether in powerful (and not so powerful) governments or corporations who compete ruthlessly for limited resources. Even in Canada, one of the most peaceful, prosperous and wealthy countries in the world, the province of Quebec is trying to convince its people (not all its people, just perhaps the French-speaking ones) to separate from the rest of Canada.

As a market economy evolves, competing in the market for a better share of the market or securing a dominant position will not be the only issue. The real competition will be in resources, aside from competing for the market to sell products and services. The real concern for corporate managers will be the resources shortages which will occur in the foreseeable future. The days of collecting all the honey and hiding it in the bee hives may not be too far off. (See Table 7.2.)

Table 7.2 A unifying theory approach to corporate vision development

Discipline	Corporate vision (through and formatted by its managers)
Natural sciences	
Physics and appreciation of how nature works. Both for new discoveries	New discoveries, source of creation and innovation
Biochemistry, biology and genetics and the protection of the environment	Substance for new product and new discoveries
Ecosystems	Environmental protection, thus to sustain life-support system for the living
Humanities	Insight of global trend in trade and economic development

A Corporate Vision

A corporate vision that is committed to the development of the mindset of individuals to create wealth and value should let the corporation be seen as a responsible corporate citizen, and state specifically its role in:

1. The protection of the natural environment.
2. Effectively and efficiently allocating limited resources and managing them responsibly.
3. Caring for people.

The reward which the corporation receives from its commitment to the above is consumers' recognition, and the knowledge that the company is the most environmentally conscious producer making the highest quality product. As a corporate citizen, it will waste no resources and make it known to the public that the entity is a community of entrepreneurs dedicated to creating wealth and adding value both for the individual and society.

8

Corporate purpose and its philosophy

Geographers tell us, as human beings, we have two choices: change ourselves to fit the environment; or change the environment to suit our needs.

Corporate managers also have two choices: embrace the theory of the firm and strive for profit, thus making the entity "entrepreneurless"; or create wealth and add value, thus building the organization, as a community of entrepreneurs.

The Corporate Purpose

The public relations (PR) manager of a large computer hardware and electronic product manufacturing company made a presentation to a group of foreign visitors about his company. It was a video presentation, beginning with the corporate purpose in big, bold letters:

1. Earn profit for our stockholders.

2. Provide for and satisfy our customers' business communication and related needs.

3. Provide quality products.

4. Provide our employees with adequate compensation appropriate to their performance.

5. Be part of our community and participate in community projects.

Immediately after the presentation one visitor asked the PR manager why profit would come before customer service. Without hesitation the PR manager answered: "Why not? That's why we are in business." After pausing for a moment, he amended his statement to say "Well, in reality they are both important, but I would say that they are like the 'chicken and the egg'; it would be difficult to say which should come first." Ironically, the company was in the midst of implementing a "Total customer satisfaction is the only objective which can ensure stable and continuous business increase" policy. Many other posters were scattered around, but the word "profit" was nowhere to be found.

The response of the PR manager regarding profit is not an uncommon one in business. Nevertheless, some members of the business community have raised serious questions about the corporation's role in society, particularly its social responsibility and ethics. The idea of focusing on ROI in a bottomline approach can no longer be taken for granted, as the narrow goal of self-interested profit-seeking is being challenged even within the corporate community itself. It was once noted by the philosopher Alfred North Whitehead that: "The test of a great society is its ability to produce good roles for its businessmen" (Halal, 1985, p. 244). In this spirit, there has been a sense of need created in the business community towards generating a more stirring vision based on human values which can inspire both institutional and societal action. Even corporate managers and executives, whom you would expect to be the most entrenched in the ROI concept, perceive this need, as relying on profit alone can be disastrous to corporate well-being.

One example of this is the tragic case involving silicon breast implants and the American company Dow Corning. Prosthetic implants are valuable tools in helping radical mastectomy patients overcome the trauma of their surgery; perhaps unfortunately they have also become a major part of the multi-million dollar cosmetic surgery industry. Recently, documented cases of prosthetic breakdown and the sometimes horrendous problems experienced by implant recipients have resulted in damage lawsuits and extensive media publicity. One case alone involved a jury awarding the victim US$25 million. The result was a cascade of legal actions, all with the basic thrust that Dow Corning was negligent in their continued production and marketing of the harmful implants. While Dow Corning successfully defended their case in a number of situations, negligence had been proven, and it would only be a matter of time before they would be buried by the fallout. A compensation package was prepared for American women who had received the implants, but it was not enough to save the company from ruin; the company executives filed for bankruptcy (though curiously enough, in the US a company can still continue to do business, even while in a state of bankruptcy). The drama is not over though; lawsuits continue to be filed against the parent company, Corning Chemical. Unfortunately there are no winners in these cases; not Dow Corning, and certainly not the women who received the implants in the first place. The only real winners are the lawyers who will take on any case as long as there is a possibility of a 40% interest in million dollar lawsuits.

The unfortunate Dow Corning incident reminds me of the famous Russian artist Wassily Kandinsky (1866–1944), who abandoned a career in law and studied art in Munich. He embedded the experiences of his native tradition of religious iconography in his work rather than adopting the European tradition, and established a new artistic paradigm. Equating naturalism and materialism, he believed that it was the mission of the artist to instil a consciousness of the soul of man and the hidden laws of the universe into a materialistic society which was otherwise doomed to destruction (*Techniques of Great Masters of Art*, 1985). I am sure that if Kandinsky were alive today he would tell Dow Corning executives who tampered with nature for profit: "Told you so."

Although "profit" is not humanity's Messiah, it is unarguably the motherhood of business. This is what makes the gravity of the situation so fundamental; it is not the ideological argument of the significance of profit, but the ambiguity of the idea itself. "Profit" as we know it has no real substance outside of the minds of accountants and microeconomists. The fundamental concept has no foundation itself. If we could free ourselves from the bondage of their ideas and learn to appreciate what the realities are, we will soon discover that:

1. everything we can reach in the universe is a form of energy;
2. we cannot create or destroy energy, but
3. we can direct energy flow, and through human efforts,

4. use the resources (energy) that the earth provides for us, to

5. increase the utility function (add value) for humans.

Our economic endeavour, including all trade and servicing and production of everything from handicrafts to silicon computer chips, must be intended to create and innovate. The purpose of a corporation is to produce through collective human effort a different combination of resources to increase their utility function, and add value with respect to people's needs. It is hard in this light to appreciate what microeconomists have in mind when they talk about "profit". The whole idea of making a profit is a betrayal of the concepts of Adam Smith, whose philosophy they claim as the source of their inspiration. *The Wealth of Nations* is a summation of his ideas on the market economy, ideas which existed long before his birth, but which he can be fully credited with for formalizing and developing into an economic doctrine. In no way can it be considered the "origin of capitalism", and no doubt he would be horrified to see how it has been used as an excuse to grab wealth and line people's pockets.

The contrast between the notion of profit advocated in the corporate environment and the value-added concept is mirrored in the real difference between managing an entity under the theory of the firm, and the development of a corporation under the auspices of the ideology of entrepreneurism. The former requires everyone in the entity to be part of the big profit-making machine ultimately for the benefit of the investors and top executives, whereas the latter is a collective effort from a community of entrepreneurs working towards adding value to society as a whole.

This idea was put forward to a group of approximately 250 mature business students. The following two comments summarize their general reactions:

"This is so unreal; how can you replace 'profit' with 'value'?"

"To ask corporate managers not to think about and grasp for profit, and instead to think of pursuing 'adding value' for the common good of society? Maybe in a dream, but not in the real world."

Perhaps they are right; perhaps it is a dream. But on the other hand, virtually all of us know of Christ's teachings, and the Ten Commandments of the Bible. They are a simple text of less than

100 words, yet they form the moral basis of the entire Western civilization, and establish the fundamentals on which basic Western laws are founded. One commandment reads: thou shall not covet thy neighbour's wife [or husband] (you may note that the Commandment has just been amended to reflect reality). To some people, this is a laugh! Thou shall not covet thy neighbour's wife (or husband) is just not real in this world. There are many Hollywood movies about wives cheating on their husbands and vice versa. Meanwhile the same stories are happening in the actors' real lives, as in the lives of countless others. And those who do not cheat on their husbands or wives often cheat their business partners. The justification is often the same as well: "Everybody else is doing it", or "after all I've done, he or she doesn't deserve any better!" Nevertheless, we still have not abandoned the dream of the "Ten Commandments". Imagine what would happen if we did. By the same token, we should never abandon the concept of adding value, no matter what anyone else is doing. Of course, there is always the question that if a corporation is not in the business of making profit for its stockholders, why would anyone be interested in investing in it? This topic shall be dealt with in later deliberations.

Philosophy and the Corporation

In simple terms, philosophy is the pursuit of the wisdom of things and their causes, whether they be theoretical or practical. The existence of a corporation, from the point of view of society, should be the pursuit of a practical wisdom, and the purpose of corporate management is to create and maintain the corporation as a community of entrepreneurs. This involves serving the needs of people, enhancing individual skills, knowledge, and personal values, and caring for the well-being of both the individual and the community. In functional terms, this wisdom may embrace the following:

1. Serving people serving themselves.
2. Serving people serving others.
3. Serving people serving our environment.

4. Serving people serving those who are unable to help themselves.

5. Serving future humanity.

In order to be clear in the intent, serving differs from helping. When a person joins the military to serve a nation, helping is definitely not the right word. They differ in cause and motives; serving is a clear obligation and responsibility, while helping is driven by different motives, including, among other things, passion, love, sympathy and caring. It must be clear that a corporation's role in society is to serve and not to help, and should be perceived as such by all concerned.

An individually focused corporate philosophy

Each individual has his or her own philosophy in life, and they are all different. Some would be considered to be noble, others would not. But all of these philosophical ideas are motivators and have their practical value, and the same holds for corporate philosophy. A corporate philosophy based on the idea of serving the people within the organization is only relevant and meaningful to them. If the pursuit of profit in a corporate structure is the requirement imposed on every individual, then this serves only the interest of stockholders and top executives who have incentive bonuses, and thus is only relevant and meaningful to them. If, on the other hand, the corporation has a commitment to serve all people and the environment which is vital to them, then this commitment and this corporate philosophy is relevant and meaningful to everyone. It is this metaphysics which must be borne in mind by all members of the corporate community. (See Figure 8.1.)

Figure 8.1   Individual in the corporation and corporate philosophy

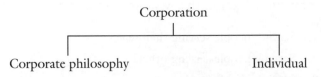

If the corporate philosophy illustrated is to be meaningful, effective and support the corporation's purpose, it must be in everyone's mindset. However, it is possible that people in the organization will change their mindset and their behaviour to be in congruence with this "serving others" corporate philosophy (as illustrated in Figure 8.2). This is not an easy question to answer. There are at least two major difficulties:

1. People will not change their mindset or behaviour to suit the corporation's needs or wants, unless by so doing they are convinced that the change will suit their needs.

2. The emphasis in Western society has been that people's needs are best suited by attaining financial wealth. Capitalism has been upheld, especially over the last decade or so, as the most effective means to attain this wealth. The general idea is that so long as you have money, you can make the world revolve around you. A philosophy of serving people will be tacitly agreed to be good for the public image, but it will not work for people who seek instant gratification of their materialistic needs and desires.

Corporate Philosophy and the Matter of Ownership

"The family that prays together, stays together." This is a motto that the author saw on a billboard sponsored by a major religious group. Those who manage corporations have long been an advocate of this saying, both implicitly and explicitly. Company-sponsored sports teams, the annual Christmas party, community projects and social events are all designed to increase the identification of the employees and their families with the company. These ideas

Figure 8.2 Individual's mindset and corporate philosophy

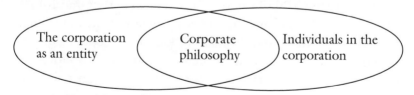

work at times, but all too often the relationship between management and the workers is confrontational. Managers complain that workers are unmotivated or "unmotivatable," while the workers hide behind the protection of the unions because they cannot trust management to work in their best interests. The struggle between management and the unions is constant, and the bigger a company gets, the stronger the union opposing it is (it has to be, or it will be crushed). Is a true resolution of this battle possible? Not unless they can find a common ground and become a true family. Could the corporate philosophy of serving people do it? Sadly no, for in a materialistic world, the common ground will have to be something material. The answer must lie in ownership. Ownership is the fundamental expression of self-interest. Of course, employee participation in ownership is probably as old as the first stirring of economic thought. Today, we have countless forms of ownership sharing in business; the idea is popular in theory, but has not been effectively employed in practice. The challenge is to revitalize this old horse before she is put to pasture.

Western nations talk endlessly about human rights. A fundamental human right is the right of decision-making, and ultimately decision-making is only possible with ownership. Today, both in the East and West, hot (real wars) and cold wars (akin to wars) are fought in the struggle for ownership. In Canada, the cries for independence for the Francophone province of Quebec have echoed louder and louder since before René Lévesque formed his separatist party. To preserve their heritage and culture the Perquistes, as they are called, feel they must have unconditional control of their destiny, not as one equal province among many, but as a sole proprietor of all their people – francophone, anglophone and others – and their land. This is ultimately about ownership. In Hong Kong, the people have recently been through a transition of ownership from Britain to China, and a struggle for the right to decide their own destiny. Taiwan under President Lee Teng-hui is in a constant struggle to maintain its identity against mainland China. The Serbs, Croatians, and Bosnians in Eastern Europe, Palestinians in the Middle East, Nelson Mandela and the black South Africans and many others have all waged famous struggles for the right to make decisions for themselves. Some of these have been more successful than others, some have conducted themselves with more dignity and morality than others, but they all share the common bond of fighting for recognition of their rights of ownership for themselves, their land and their heritage. To participate in corporate ownership is to be brought into the corporate family, rather than to be separated from those who call themselves employers, and be labelled as one of those who have no right to make management decisions, the employees. The question is how can we make it so that everyone can consider themselves as a true member of the family? There is, of course, no perfect solution. What will be described below is a solution that will bring the corporate philosophy of serving people into the mindset of every individual in the corporation, rather than being viewed as merely a piece of "public relations" propaganda.

An Integrated Approach to Employee Participation in Corporate Ownership

The fact of ownership participation is not new; details have been described in another of the author's publications (Kao, 1995, pp. 153–65). The following model is based on this work, but with an integrated approach involving the corporation's financial planning and cash flow projections.

Assume that a corporation intends to bring employees into the big family so they will feel a sense of belonging, rather than being viewed as outsiders. The executives (managers) have decided to propose to the board of directors a plan to involve all the employees in the company. The company's name is ENTREP Co. Inc.

At the previous board meeting, it was resolved that an additional two million shares of common stock would be released into the market as part of a corporate refinancing plan. The company executives and some board members believe that in order to make the corporation an entrepreneurial company, it is necessary for the employees to feel that they are part of the big family. At the request of the CEO, 100,000 shares are to be released for employees' use, while the other 100,000 shares will be offered to the existing shareholders at 40% below market price. After the subscription date expires, the remaining shares will be offered to the public. It is anticipated that among the executives who held shares of the company, other active members of the board and the shareholders, enough will subscribe to the new issue at the price stated for the company to anticipate an immediate incoming cash flow of US$1 billion. The remainder of the required refinancing is expected to be raised from the employees, but only with an acceptable plan for their participation. The total financing required from the company's growth plan is US$2 billion. The following details are assumed:

- Authorized shares of common stock of no-par value 1 million.
- Current trading price per share US$200.00.
- Shares subscribed and paid for 500,000.
- Annual sales US$800 million.
- Sales/equity (paid) eight times.
- The rate of value added to common shares calculated on the accounting method acceptable to the stock analysis is approximately 20%.
- Bank loan interest rate on a long-term basis: 12% less income tax (assumed to be 50% on corporate profit) of 6% = 6%.

Before submitting this to the board, the CEO requested the financial controller to prepare a brief cash analysis and financial performance for a ten-year period (5 + 5, go back 5 and plan for future 5) in a graphic form. These are shown in Figures 8.3, 8.4 and 8.5.

According to the above, the company at this moment is in a favourable cash flow position, and earnings are sufficient to meet all financial requirements. However, in view of the refinancing and employee participation, and assuming that all participative stockholders for the new issue will pay for their subscriptions by cash, the company's cash flow position will be further improved, but the equity and long-term debt ratio will be changed from last year's 1 : 1.5 (long-term debt US$1 billion, equity US$1.5+ billion, retained earnings US$100 million). After refinancing the debt, equity ratio will be 1 : 2+ (debt US$1.2 billion, equity US$1.24 billion, added value US$600 million, plus new subscription US$240 million. The company's assets will likely be also increased

Figure 8.3 ENTREP Co. Inc.: Breakeven analysis

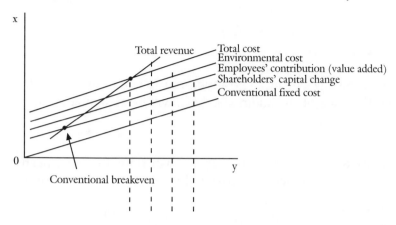

Total revenue

Total cost
Environmental cost
Employees' contribution (value added)
Shareholders' capital change
Conventional fixed cost

Conventional breakeven

Figure 8.4 ENTREP Co. Inc.: Brief financial performance and expected performance

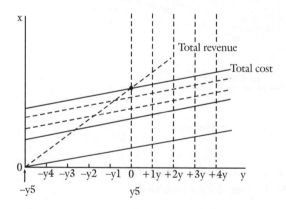

Total revenue

Total cost

by US$500 million, represented by US$30 million cash and securities, and US$200 million receivables and other assets.

This situation leaves the company in a very favourable position to negotiate a bank loan to finance the employee stock participation plan, with an after-tax effective cost of capital of 6%. If, on the other hand, the company could maintain its rate of return on equity of approximately 20%, the leverage from the borrowing would be a benefit both to the company and employees. The net result:

- The company is able to borrow from the bank to finance employee share participation.
- The company's cash flow position is improved.
- Funds are now available for expansion.

The company's financial position prior to and in the year following the issue of the additional stock will be as shown in Figures 8.6 and 8.7 respectively.

Figure 8.5   **ENTREP Co. Inc.: Cash flow, financial performance and expected performance illustration**

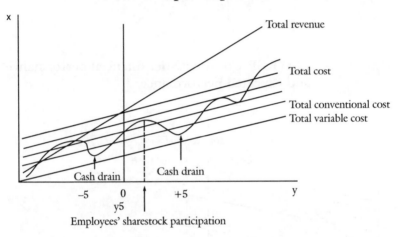

Figure 8.6   **ENTREP Co. Inc.'s financial position as at 31 December 1999 (in '000)**

Assets		Liabilities		
Cash	1,000	Current liabilities		21,000
		Long-term liability		80,000
Receivables	10,000	Equity		
		Common shares 100,000		
Plant assets	240,000	Added value 50,000		150,000
Total	251,000	Total		251,000

Figure 8.7 ENTREP Co. Inc.'s financial position as at 31 December 2000 (in '000)

Assets		Liabilities	
Cash	1,000	Current liabilities	31,000
Other investments	30,000	Long-term debt*	90,000
Receivables	30,000	Equity	
		**Common stock 124,000	
		Added value 60,000	184,000
Plant	244,000		
Total	305,000	Total	305,000

* Intended to replace profit or retained earnings, details will be dealt with in a later chapter.

** Including a long-standing agreement (approximately ten years) of employees' equity participation in the amount of US$12 million. The company may use this privilege to draw from the established credit as needed.

Please note that the above statements are prepared only to illustrate the essential idea of incorporating the employee's share participation plan into financial planning and vice versa. Only essential figures are listed in the statements. From an accounting point of view, this is incomplete, but should be sufficient to illustrate the necessary concepts.

The company may impose other stipulations to ensure that employees will not be overly burdened by periodic payment schemes and to keep the ownership within the company. The following are highlights of possible stipulations:

1. Under the employees' stock participation plan, no individual may subscribe more than 40% of his or her gross salary.

2. The participation of company stock will be financed by the company through a special loan arranged with the bank at an annual interest of 12%; participating employees are required to bear an interest of 6%, while the other 6% will be paid by the company.

3. Employee owned stock is non-transferable to outsiders; in the event that an employee wishes to part with his or her holdings, the share stock may be offered to other employees of the company at a price determined by the company on the basis of share trading price. The 40% rule does not apply to the transfer stock. However, the transfer of stock must be approved by the company,

and any one employee may not hold more than 200% of his or her annual salaries share stock under the scheme.

4. If an employee departs from the company for any reason, the share stock must be sold to the company as described in item (3). If there are no takers of the offer, the share stock should be kept in the company under the guidance of the company's treasurer for further release to new employees.

5. The company may repeat the offer to employees as deemed appropriate and necessary.

6. The implementation of the company's employees' share stock participation plan is subject to the governance of statutory requirements.

Share participation would provide all members of the corporation with a feeling of commonality and commitment to the well-being of the corporation; meanwhile the "serving people" philosophy would be more effective to obtain employee commitment, because it will be supported by "substance". As such, the plan has at least two distinct characteristics:

1. Share stock will always remain in the company, and it is unlikely that the shares would be devalued because of the scheme.

2. The plan will only be implemented when the company is in a sound financial position, strong in cash flow and with negotiating powers with the bank. As in most cases, financial institutions tend to render their support to corporations who have a sound financial position, good human resources policy and potential to grow.

Figure 8.8 suggests that through the reinforcement of stock participation, individuals in the corporation will further strengthen their relationship with it, and adhere to the corporate serving philosophy because it has substance rather than just empty words. The consequences are easy to predict; individuals, in participating in ownership, become part of the team and share the same

Figure 8.8 **Employees' involvement in corporate philosophy through ownership participation**

Individual ownership +/− participation = Corporation philosophy

philosophy. Some years ago, a corporate group disenchanted with government intervention in business spent a large sum of money on a full page advertisement with a bird and a simple message: "Free the enterprise", now it is time for corporate decision-makers to think about the fundamental human right in corporate structure: "Give them the right to own. A corporation is an entity of people, not a computer or machine." (See Figures 8.9 and 8.10.)

Figure 8.9   Individuals in a corporate without equity

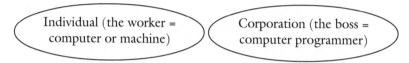

Figure 8.10   Relationship between the individual, stock owners and corporate philosophy

The "Quality" of Partial Ownership

There have been many types of employees' participation in corporate ownership. Some of them are bonus type giveaway plans to award executives for their superior performance. Others use different types of share plans including preferred, common stock A or B, with voting power, or without voting power. All of these limit employee participation to minority positions. In other words, corporate control is still in a major or block stockholder's hands. This means that if ownership means decision-making, employees who participate in the plan have little influence on major corporate policy. This is true even in countries which have legislation defending minority stockholders' rights, as major decisions still lie very much in the hands of key individuals who either hold majority or block shares, or are on the board, in particular as CEOs.

The bottomline for ownership and decision-making is that, even with a fraction of the total shares, an owner is much better off than not having any shares at all. Quality of ownership does not imply that once a person holds one corporate share, he or she will be interfering with the day-to-day operation of the corporation; rather, it is a matter of how corporate democratic rights are to be exercised. In numerous corporations where employees are also shareholders, they do not normally exercise their participatory right to any great extent. For example, before every shareholders' meeting, the management normally makes a report outlining key activities during the past year and financial information requesting the presence of all shareholders to attend the meeting to vote on major issues recommended by the board. Minority shareholders seldom exercise such rights, and consequently the board of directors typically has free rein, even when technically they do not have majority control. It might be argued that a corporation that is prepared to accept employee involvement in decision-making would do just as well to form a management committee or elect employee representations to the board of directors; this is done in some American corporations. Why bother with an employee participation scheme? While the schemes tend to involve employees to some extent, there is a clear difference between the two. Quality of ownership implies that the owner must treasure ownership rights and accept ownership responsibilities, which is not possible under the other schemes.

The Corporate Philosophy and Ownership

A corporate "serving people" philosophy is both prescriptive and descriptive. It is prescriptive because it guides what the corporation should do, and descriptive because it outlines what the corporation is in fact doing. It is not necessary to be a part or full owner to acquire the "serving people" philosophy; as a prescriptive philosophy it is about perception, and a perception is not so much a reality, as a commitment to a reality in the making. Here are some suggestions:

1. Include the corporate serving people philosophy in the prospectus inviting employees into the share participation scheme.

2. Highlight the importance of corporate philosophy to the company's competitive position, and such a philosophy is more than just market-driven or customer-driven. It does not need to be elaborated.

3. Explain to potential share subscribers, the meaning of serving people, not as a slogan, but in plain language in no more than a two-paragraph personalized letter to employees from the CEO or board of directors.

4. Explain what the corporate democratic rights are and how they relate to corporate philosophy in plain and simple language. Avoid using it as a propaganda piece to enhance the corporation's public relations image.

9

Corporate value

"Value" is a personal treasure and property, and it is not for sale. Consequently, to impose a "foreign value" on an individual can be meaningless as well as useless.

A corporation is a community made up of individuals, and corporate values should not be something perceived only by the corporation's CEO, the board, any single individual or interested group, but should be an aggregate of values possessed by every individual in the organization.

From Corporate Philosophy to Corporate Value

Despite the existence of hardcore people who still believe in "profit as the supreme purpose" for the existence of corporations, the emergence of the concept of the human economy[1] has made corporate managers take a second look at their existence and purpose, just when the world is about to confront a new century. With the twenty-first century will come increased automation, computerization, a growing information highway, and new scientific breakthroughs. On the other hand, it will also bring failing ecological systems, and a growing need to serve people while demanding a better and more sensitive approach to the protection of the environment. The most exciting moment of the new century will be the day when we can all witness the free flow of goods, services and labour. When this day comes, we will tell ourselves: "We, the people, are and always will be, the ones who really make a difference."

Without intending to steer away from the "capital-driven" focus of management in a corporation, people's needs and interests must receive more attention from corporate managers, in particular, the

174

needs and interests of those working in the organization. In as much as management is in the decision-making or command position, the true nature of management is no other than to serve the corporation, including the corporation's employees. When the employees are treated well, they in turn will serve others in the market well. It is the philosophy to serve people that prompts corporate growth.

There is a little story about a fish and the water. Once a fish told the water: "It is because of me that you are alive and vibrant. Without me, you are nothing but a pool of filthy and lifeless water." The water quietly replied: "Yes, my dear fish, you make my life interesting, colourful and vibrant, and if you leave me, my life will not be as exciting though I will still live. But if I should leave you, you surely will die." The same may be true in a corporation's environment. Although management is important, without employees, what is there to manage? The introduction of the "serving people" corporate philosophy is really just a fresh face for an old wisdom that the basis of any civilization or culture is its value system.

Corporate value – made to order or "order to take"?

If serving people is the philosophy of a corporation, then the corporate value system must be the spirit that guides the individual to act for the individual's interest and the interest of the corporation, and thereby, of society. Building corporate values is not a simple task, unless, of course, the value system represents only the proprietor's personal values (order to take, so to speak), with very little input from those below the proprietor of the firm. Simple, but not effective; in a corporate environment with a large number of people involved, if the value is to be accepted by everyone in the corporation, they must be involved in and be part of the value system.

How does a corporation perceive and build its own value? Perhaps it is useful to borrow from the wisdom of democracy and let all members of the corporation offer their own values, aggregating them into a single value rather than taking the traditional approach of pushing a value system down the employees' throats. It is the wisdom of democracy that participation makes people feel a sense of belonging, and gives them the coherence of the spirit of the value system they have built. The challenge is how to assemble the values of all individuals and congregate them into a single system.

To aggregate things that cannot be aggregated, the challenge of finding a common ground

Economists have been troubled by their inability to aggregate economic variables. Although accountants try to convince the world that it is only the monetary unit through the function of the market system that is able to consolidate everything into a common denominator, this does not particularly meet with everyone's satisfaction. As some business person once remarked:

> Economists can talk about profit (in real terms), but they cannot show us what profit is. On the other hand, accountants can show us all kinds of profit (through the common expression of the monetary unit), but we don't believe them.

Ironically, we are not happy about the use of the monetary unit as a substitute for aggregate economic goods and services, yet, thus far there is no alternative. It works as well as it does because the use of the monetary unit represents an aggregate which is unbiased, through the magic words of supply and demand. Thanks to the accountants, it is possible to find a common ground even though it is illogical to aggregate a computer, a theatre ticket, dish washing detergent, or cosmetics with chicken liver or blood sausages, but in monetary units they have done so (Figure 9.1).

It may be worthwhile for us to consider the two-sided nature of the market economy. It is a blessing from heaven since without it economists would not be able to tell us what to do, and accountants would not be able to tell us how to account and what to account for. Under the circumstances, we would have no need for "B" school professors, business persons, corporate managers, and certainly there would not be any reason to write about entrepreneurship or an entrepreneurial approach to corporate management. In fact, we probably would still be living in caves, much the same as our monkey ancestors, apes or monkey-like animals. Yet

Figure 9.1 Economic variables aggregated under the market system: Governed by supply and demand

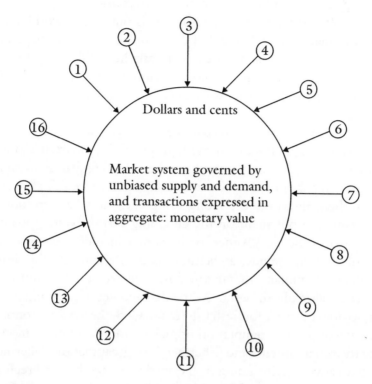

Dollars and cents

Market system governed by unbiased supply and demand, and transactions expressed in aggregate: monetary value

Legend:

1. Timber
2. Clothing
3. Eggs
4. Fish
5. Home decorating
6. Taxi service
7. Computer software
8. Printing
9. Liver sausage
10. Tableware
11. Cosmetics
12. Detergent
13. Toys
14. Courier
15. Wafer fabrication
16. All others

the market economy can be quite evil, and almost unreachable, because it is armed with the touchable unbiased price system and is governed by the irresistible forces of supply and demand. We really should have a closer look at the market economy just to know what it could do to us, and we shall do so, for the moment, through the eyes of a fish.

Hong Kong is the fastest economic growth region in the world, and it has food and restaurants to match its high-rolling status. Hong Kong restaurants specialize in rare fresh and live fish. For example, the lips of a blue parrot fish are considered a prime delicacy; soft and tender with a taste that is out of this world and a price to match: it fetches US$300. To the rich Hong Kong people (no one seems to take the trouble to ask: how did they get to be so rich?) who can make millions by just making a few "deals", what is US$300? In the market economy, money, to the rich is easy come, easy go.

As people are willing to pay the price, on the demand side it motivates restauranteurs to pay high prices for the fresh and live catch. Hence, it prompts the fishermen (big companies or small fishermen alike) to use every means to catch those fish alive and keep them alive until the Hong Kong chef can take them out of the tank and put them on the chopping board. Netting is only one part of the story, other methods may include the use of the deadly poison cyanide, a chemical used to administer the death sentence to humans, that in low doses will drug the fish; unfortunately, it also kills the coral. As the demand gets higher and higher, so does the price for fresh fish (thank God, the market economy really works). The temptation of making more and more money for the restauranteurs and fishermen is so great that the fishermen will do anything, including ruthlessly destroying the coral reefs to take the last surviving fish. Let's give thanks to the glorious, most powerful and unbiased price system and the great Marshallian theory of supply and demand that have really made Hong Kong the "pearl of the east". Believe it or not, the fish stock and coral reefs are really like the lily pond story[2] on their twenty-ninth day on earth. Soon enough, there will be no more fish, or coral reefs, and what will be left is a dead seabed.

Governments are concerned about what is happening. For one, the Philippines, which is the main source of the beautiful marine fish that satisfy the rich Hong Kong consumers, is quite serious about dealing with illegal fishing methods (such as using cyanide), but fishermen, motivated by profit, continue to risk taking fish from the same spots and at the same time moving further south to meet the heavy demand from Hong Kong. The issue is obvious, and the case is a clear reflection of how the market economy works.

What matters most is the Hong Kong Government's stand. Obviously, to take legal action against the fishermen is ineffective. The other possibility is to ask consumers not to eat these endangered fish, which is, of course, impossible, because under the market economy, to question the wisdom of rich consumers is absurd. There is only one way that will help and it is for the Hong Kong Government to take a stand and ban all restaurants from selling the endangered fish species. Can it be done? Yes. But will the Government do it? Not a chance, at least at the moment, because this is part of Hong Kong's growth story. How can it be possible for anyone to hinder Hong Kong's growth? The last British Governor of Hong Kong, Chris Patten, said to the public: "I would rather be part of the growth than be part of a recession" (these are not his exact words, the author has paraphrased a televised statement). Would the Hong Kong Government ban the eating and selling of endangered fish? Under the circumstances, it is almost unthinkable. After all, we are only talking about fish which cannot put up a poster or grant reporters an interview. However, there is one thing that fish can do: die.

Hong Kong's prosperity has always been a mystery. In the early days, it benefitted from the emissaries of China. As China turns its wheel to a "market economy" and because Taiwan and China are not directly (at least officially) on talking terms, everything to do with Taiwan and China would have to go through Hong Kong, which means money for Hong Kong. Making money from crises occurring in China is not enough, now it is exploiting the environment, in this case, the fish. Some people might say, "Well, so what, fish do not even talk!"

Putting aside Hong Kong's gourmet fish story, let us get back to how to make corporate value an aggregate. Aggregation, logically, should be of things of the same kind. Hence, at first glance, we cannot logically aggregate family honour with the Ten Commandments, or lump "kindness to animals" with "do not cheat on your wife or husband". Yet, is it possible to aggregate all values into a single term which we normally call: "decency" or the "common good"? There is a North American expression that says it all:

"Motherhood". Let us see if we can find some comparable motherhood words.

Corporate Value, Can We Find a Common Ground?

Any attempt to aggregate all individuals' values into a single form of corporate value can be just as impossible as to try to put various resources, goods and services into an aggregate. There is a need to find a common ground which will make everyone within the organization feel that he or she is part of the organization, with a shared value belonging to each one of them. The author recalls the time when some concerned people all over the world, particularly in the United States, were incubating the idea of creating a world body named the United Nations. The idea was for human beings to learn to settle their differences at the discussion table rather than on the battlefield. We were troubled as it was not easy to find a common belief or idea that would weld the nations together and realize the dream. Accordingly, a great deal of human effort was put into the search for answers to two important questions (just a few years before 1945):

1. Why did the League of Nations fail to achieve its purpose?
2. Looking into the wisdom of cultures, religions and values of people in every country in the world, is it possible to find a common belief and trust that will bring people together under the name of the United Nations?

By so doing, it was possible to create a body with a value (or set of values) which we could all believe in and share. One of the most important of such values was human rights. Yet, even though the Human Rights Charter was signed by all member nations, after some 50 years, all the challenges remain, much like: "The old soldier who never dies"; unfortunately, they do not even fade away.

Similarly, the United States is a single nation with 50 states. They share one constitution, a common currency and one language, but have different tax laws, and a variety of legislative differences, even to the extent of the use of capital punishment. The European Union (EU) may not be a perfect model for European unity, as proclaimed by the German Chancellor, Helmut Kohl: "Germany is the fatherland; the European Union is the future." Certainly the EU model cannot be compared with the established US model, but it has reached the stage where there is a distinct possibility that Europeans (I mean all Europeans) can perceive themselves as citizens of one big European family. These are the aggregates: The United Nations, the United States and the European Union. Recently, ASEAN (Association of Southeast Asian Nations) created their own perceived common value, though they have a long way to go compared with the Europeans. Nevertheless, they all attempt to have a "common ground" to unite people for the purpose of being alive and living.

All values must be based on some acknowledgeable elements for the common good. In a corporate structure, "values" from individuals could be aggregated into a single value system under the roof of the corporation. The corporate value would include values of all individuals but without individual identity. It could be the same as the "melting pot" idea of American culture; a nation claiming to have integrated hundreds of different races and cultural origins and melted them into a big pot. To do the same in a corporation is a difficult task. Certainly, a simple and easy answer cannot be found in this book, but by beginning to gather different values from possible sources it would help the development of a meaningful common value. It could be shared by everyone in the organization, while establishing a direct link with the corporation's serving people philosophy.

A Short Visit to Values of Different Origins

There has been much talk and excitement in respect to Japanese culture and the Japanese business value system that has contributed to Japan's economic success. To know Japanese values is a great challenge, and we may have to trace Japanese history back to 1,000 years before Christ.[3] For illustration purposes, a few noted corporate values along with individuals' values could be summarized as follows:

1. Unquestioning devotion to the national entity. Japanese businesses were confronted with serious problems during the recession period, but the devotion of Japanese business persons to the national entity drove them to maintain and improve employment at home, rather than laying off people just for the sake of maintaining profit which often occurs in Western business entities.

2. A unique corporate commitment to their employees. The corporations place employees as their "first priority" of importance, then come their customers, and last, their shareholders – just the opposite to the US corporate mentality.

3. Shintoism is the preconditional energy for refinement of business ethics: the feeling of loyalty to the Japanese practice.

4. Mental stability; concentration during work.

5. Self-fulfilment.

In Islamic countries, family honour is an unquestionable value that governs everyone's conduct and behaviour. For example, if the actions of a family member are deemed a dishonour to the family by family members, the person who committed such dishonour could be killed by them. The honour system embedded in the society and the "value" of such a system as perceived and observed by people in these countries have made the countries as they are.

Christian value has been a treasure to Westerners and to those who believe in Christ throughout the past twenty centuries. In Asia, the mass majority of people may have inherited their values from the Confucius doctrine, other philosophers and great teachers, or from their religions or both. At times, political leaders also exemplified the values of their people through their personal values; these leaders include Abraham Lincoln, the 16th President of the United States, and Zhou En-lai, the late premier of China. Lincoln imparted his values on liberty, not only to Americans, but to everyone who believes in the meaning of "freedom". The legendary wisdom "of the people, by the people and for the people" remains beyond compare. His values and contributions to the individual's right remain in the minds of billions, not just as memories, but as a guide both in political thoughts and individual actions. Zhou En-lai's unselfishness and lifelong struggle for people in

China earned him the United Nations' respect. On his death, flags in front of the UN building were at half mast. When some member nation representatives to the United Nations asked the Secretary-General the reason for lowering the flags, he responded: "Can you tell me of any leader in the world who served the people all his life and died without any tangible wealth or a single property in his name?" The greatness of his value given to this world is not anything that he said or wrote, but the "living" example of his personal dedication which earned him respect from the world body and people globally alike. It was Zhou En-lai and then US President Nixon's envoy, Henry Kissinger, who made the first contact to bring China out of seclusion to be reunited for peace and prosperity with the rest of the world.

Of course there are almost infinite doctrines and religious teachings, whose good deeds could all be turned into individual personal values. For example, followers of Buddhism generally believe in doing good, promoting the fundamental value of mercy and kindness towards all living beings and advocating punishment for wrongdoing; Confucian social conventions are followed not only by people in China, Korea and Japan, but in other countries as well.

More of Chinese Cultural Values

China is a nation with a long cultural history. Kirby and Fan (1995) made a list of Chinese cultural values, some of which are very much embedded in the individual Chinese mind, and resemble some Western thoughts as well. They are:

NATIONAL TRAITS

1. Patriotism
2. A sense of cultural superiority
3. Respect for tradition
4. Perseverance
5. Knowledge (education)
6. Governing by leader instead of by law
7. Equality/equalitarianism

8. Trustworthiness
9. *Jen-ai* / kindness (forgiveness, compassion)
10. *Li* / propriety
11. Tolerance of others
12. Harmony with others
13. Courtesy
14. Humbleness (modesty)
15. Observation of rites and social rituals
16. Reciprocation of greetings, favours and gifts
17. Repayment of both the good and the evil that another person has caused you
18. Face (protecting, giving, gaining and losing)

SOCIAL (FAMILY) ORIENTATION

19. Filial piety
20. Chastity in women
21. Loyalty to superiors
22. Deference to age
23. Deference to authority
24. Hierarchical relationships by status and observing this order
25. Conformity/group orientation
26. A sense of belonging
27. Reaching consensus of compromise
28. Avoiding confrontation
29. Benevolent authority

WORK ATTITUDE

30. Industry (working hard)
31. Commitment
32. Thrift (saving)
33. Persistence (perseverance)
34. Patience
35. Prudence (carefulness)
36. Adaptability

37. Non-competition
38. Not guided by profit
39. Moderation, following a middle way
40. *Guanxi* (personal connection or networking)
41. Attaching importance to long-lasting relationship and not gains
42. Wealth
43. Resistance to corruption
44. Being conservative
45. Long-term orientation

PERSONAL TRAITS

46. *Te* (virtue, moral standard)
47. Sense of righteousness/integrity
48. Sincerity
49. Having a sense of shame
50. Wisdom/resourcefulness
51. Self-cultivation
52. Personal steadiness and stability
53. Keeping yourself disinterested and pure
54. Having few desires
55. Obligation to your family and nature

ATTITUDE TOWARDS ENVIRONMENT

56. Fatalism (believing in your own fate)
57. Contentedness with your position in life
58. Harmony with nature

It is not surprising that many of the above resemble Western thoughts, while on the other hand, quite a number are contrary to today's corporate managers' thinking and practice. Most marked of the latter are:

37. Non-competition
38. Not guided by profit
44. Being conservative
54. Having few desires

Some of the cultural values listed may not be practical in Western situations. They include:

9. *Jen-ai* / kindness (forgiveness, compassion)
39. Moderation, following a middle way
41. Attaching importance to long-lasting relationship and not gains
45. Long-term orientation

On the other hand, these are also moral issues. Whether or not they are considered as personal values is the individual's choice, although some are more universal than others; for example, it is unlikely that many people in any society will openly admit that he or she has no sense of "shame".

Perceived Differences in Values Between East and West

Human beings of this era have come closer together than ever before, especially now that the gap between East and West has narrowed. This is most true at high levels, with business executives and political leaders meeting through APEC (Asia-Pacific Economic Cooperation), and ASEM (ASEAN and European Meeting [Summit]) being cases in point. Nevertheless, there remain massive differences on issues, including trade disputes, and even more so, on individual perceived values. Table 9.1 shows the North American business executives' concept of values as compared with their counterparts' in Asia.

These differences illustrate both the need for and difficulty of finding a common value, if people are required to work and live together as a "family", especially in institutions such as multi-national corporations. (See also Table 9.2.)

Corporate Value

A corporation is an economic entity, not a human one, but human beings are the ones who make decisions for the corporation. Therefore, it is humans who decide on what the corporation can do for

Table 9.1 Values of Asian and North American executives, ranked by importance

Rank	Asia	North America
1.	Hard work	Freedom of expression
2.	Respect for learning	Personal freedom
3.	Honesty	Self-reliance
4.	Openness to new ideas	Individual rights
5.	Accountability	Hard work
6.	Self-discipline	Personal achievement
7.	Self-reliance	Thinking for yourself

Source: *Asian Wall Street Journal,* 5 March 1996, p. 11. Used with permission.

Table 9.2 Key country differences (values important to some countries but not to others)

Country (region)	Value
Japan	Harmony
Hong Kong	Orderly society, personal freedom
Singapore	Orderly society
Thailand	Achieving financial success
South Korea	Thinking for yourself
Taiwan	Self-reliance

Source: *Asian Wall Street Journal,* 5 March 1996, p. 11. Used with permission.

other human beings, and in the name of the corporation create the corporate values. There are many examples of collective wisdom that can be expressed by an individual who represents the corporation, and in the name of the corporation. For example, during the 1980s, when the steel industry was downsizing operations in the UK, British Steel, in a difficult financial situation, was willing to contribute a lump sum of money, plus rendering its human resources to help the jobless and/or individuals laid off from the industry to create smaller enterprises to provide jobs both for themselves and employment for others. The endeavour was called: "Put

Something Back" which constituted a great corporate value at the time. Similarly, in responding to a reporter's query about Singapore's 1996 budget, the Chairman of the Parliamentary Committee (Finance) remarked that it was a budget based on the government's philosophy "to return the fruits of growth to the people" (*The Straits Times*, Singapore, 2 March 1994, p. 35). True, even though some people were calling it an election budget (*ibid.*), it is the spirit of the expression: "to return the fruits of growth to the people" that make it a "value" that reflects the government's stand on people. Along the same lines, a corporation could also have a value built on the corporate's serving people philosophy.

Aside from the above, there are also prescriptive business values advocated by well-known individuals, such as J. R. Hicks (a renowned economist). According to him, if a businessman sold his machine to a manufacturer and made a profit of £500, he should not consider that he made a profit of £500 unless the manufacturer who bought the machine made goods with the machine, sold the goods and also made £500.[4]

Formation of a Corporate Value

Unlike materialistic matters, a value is not something that we can touch, feel or handle; it is intangible. Moreover, it is often not even easy to describe. It is a spirit, a thought and a commitment. In reality, it is a motivational factor or driving force that makes people do things of value for others and for the common good. To form or create a value in a corporation should be a total effort. The devotion to the formulation of a corporate value would embrace personal values of all individuals in the organization. Although it may not be earth shattering, the search for a common ground to build a corporate value is certainly no small task. There has been no clear evidence of a prescribed method to tell corporate managers how to create a corporate value; nevertheless, there are three possible approaches that could be considered.

Approach 1: The founding entrepreneur's personal value as a corporate value

This is the most common approach to formalize a corporate value. Almost all of those are found in entrepreneur managed corporations. In Canada, T. Eaton Company is a family business and an entrepreneur managed retail giant. As the founder of the corporation did not approve of smoking, it became an Eaton's store value. Throughout the generations, regardless of how tempting it may have been to sell cigarettes and provide a very high profit margin for the business, no Eaton's store has ever sold anything that has to do with smoking.

There is another case in Singapore. A large corporation CEO (founder of the company) treasures paper, believes that no one should waste paper on any account. Under the circumstances, in his company, no tissue paper is provided in the toilet, and those who are in need of tissue paper must obtain it from the supply store clerk. Of course, this may be considered to be the CEO's unusual behaviour, an attribute difficult to publicize, but a personal value. Consequently, everybody in the company has learned to save paper. Since paper comes from trees, trees are saved. Come to think about it, if cutting down a 50-year-old tree would cost our environment $194,250, it is definitely a good measure, both to save cost for the company on the one hand, and to lessen our demands on our green environment, on the other.

These forms of entrepreneur's personal value prove to be a formalized value of a corporation that are illustrated in Table 9.3.

Depending on the management style and personal relationship with the rest of the people in the organization, an individual could completely believe, work and be guided by the same set of values as those of the owner/entrepreneur.

Approach 2: A corporate value formulated through a more systematic approach

Inasmuch as a "value" is what an individual perceives, it is unlike an object or anything with substance which can be packaged, streamlined, bundled or worked with like a computer. Yet it can

Table 9.3   Entrepreneur's personal value transfer

Stage	Transferring process
1.	Venture founder/owner's personal value
2.	The same value informally communicated to others in the organization through personalized management and interaction with others
3.	Others appreciate the purpose of the value, through the working relationship and informal interactions
4.	The entrepreneur's value has a penetration effect and impacts to the individual's mind
5.	The individual appreciates the entrepreneur's value and makes it part of his or her own value system

be developed systematically. The essence of a systemized approach is how to deal with the challenge of gathering personal values in large organizations where works are compartmentalized, with fewer opportunities for people to interact with one another. However, attempts can be made, human factors can be gathered and corporate perpetrations can be assembled. This is illustrated in Figure 9.2.

The principal idea of the above approach is the inclusion of both the individual's personal key factors and the corporate need for survival and growth. They can be "melted in a big pot" of corporate value, hence, everyone in the corporation can identify with it.

Part ownership, through a corporate employee equity participation scheme, is the basis for the individual to form a sense of belonging. Reflecting personal factors in the corporate value is the spirit that welds individuals into a single corporate philosophy and purpose. The philosophy of serving people through a collective effort makes the corporation a community of entrepreneurs for the purpose of creating wealth and adding value both for the individual and the common good.

Figure 9.2 A more systemized approach to the formation of a corporate value

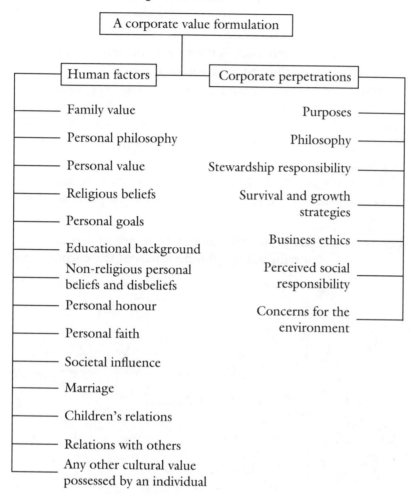

Approach 3: A corporate value derived from the collective wisdom of people

Perhaps it is best to borrow the wisdom of Abraham Lincoln: a corporation is a public entity, a stakeholder in the society, and an organization created "by the people, of the people and for the people". It is the people that the corporation serves, therefore, a corporate value should also be derived from the collective wisdom

of people to create a common purpose to which everyone contributes. It is illustrated in Figure 9.3.

It should be noted that the above approach only considers
human factors, not corporate perpetrations. This is because it is
assumed that corporate value is the same as the individuals' values
in aggregate, in contrast to the traditional approach based almost
entirely on corporate interest as perceived by key individuals.
Individuals in the traditional approach work under a contractual
agreement between those who make decisions and those simply
working according to the prescribed job description in the
employment agreement.

In short, it is perceived that individuals are likely to accept the
corporate value, if the value was created for the purpose of the

**Figure 9.3   Corporate value as a melting pot of values
            contributed by all individuals of the corporation**

People-oriented human endeavours

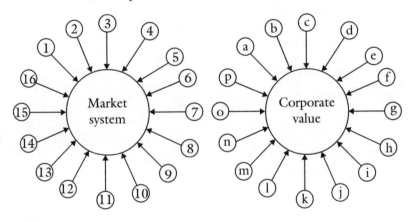

Legend:

1. Chicken liver
2. Computer software
3. Shoes
4. All others
    (see Figure 9.1)

(a) Individual's personal philosophy
(b) Family value
(c) Religious belief
(d) All others (see human
    factors in Figure 9.2)

Note: Every product or service in the market system has opposing forces: the
force of demand and the force of supply.

common good and includes part of themselves as integrated parts of the great melting pot.

Formalization of Corporate Value and Development of a Mental Model to Monitor Change in the Corporate Value Over Time

Formalization of corporate value

A corporate value is the same as an individual's personal value. The value itself can be as abstract as your perception, but it can be transformed into a more concrete form through corporate behaviour or action by humans. Take, for example, Japanese corporate values. One of the many corporate values is concern for the environment. Based on this fundamental value of concern for the environment Fuji Film came out with an advertisement saying nothing about the company. The entire advertisement was a description (an attractive video with music) of the environment and it was so beautiful that it was awarded first prize in an advertising competition. Although what Fuji Film did was for profit, it was also for more than profit. It created wealth for their employees in respect to their concern for the environment, as well as a competitive strategy. Needless to say, the whole idea was to encourage consumers to cultivate loyalty and buy Fuji films, increase the company's revenue, and maybe profit, and also add value to society. They fostered a love for birds, trees, and above all, our environment as a whole. Assuming one corporate value that has been set is "loyalty and friendship"; if the corporation truly observes its corporate value, just think of all the good deeds the corporation could do.

The difficult challenge is for corporate managers to collect all the wisdom and values from individuals, particularly, since some personal values change over time. There is no quick and easy way to do the task; however, here are some steps that may be considered:

Step 1.   Gather information about every individual's personal values through a planned job interview and job application form.

Step 2. Re-emphasize the personnel function from staff control to monitoring how the staff is being served in the corporation, including coding staff personal values from information gathered from job interviews and application forms.

Step 3. Analyze employees' personal views and identify similarities and differences.

Step 4. Highlight similarities and explain the differences.

Step 5. Consolidate findings, draft a corporate value statement and invite suggestions and comments from employees. It should be noted that this is the area where an individual could identify him or herself with the corporation, much as with a person's name or family relations, and say "it has my flesh and blood in it." One very successful entrepreneur from Indonesia remarked that the success of a company is reflected by the length of service of its employees. A good, successful entrepreneurial corporation could easily find employees working in the organization for thirty or forty years. Many of them will never retire (of their own accord). How can a company expect anyone to stay with the organization for thirty or forty years, if he or she is not part of the corporate value system?

Step 6. Finalize corporate values as the basis for development of a corporate enterprising culture.

What is the corporate value? Unfortunately, there is no list available, and it would be foolish to predetermine what it is, since it must be developed through a collective contribution from all those in the organization. But there are examples, some of which have already been noted earlier in the chapter. There are a great many others; one such is seen in the motto of De Faseo of Hamilton, Canada, a steel company: "Our product is steel, our strength is people."

If it is correctly interpreted, the company is committed to its value that people come first before anything else. But if the management decides to lay off and discharge people on account of "profit" without adequate compensation, then all we can say is that this company's management is a liar, and its corporate value is nil.

10

Corporate enterprising culture

Enterprising culture is "a commitment of the individual to the continuing pursuit of opportunities and developing an entrepreneurial endeavour to its growth potentials for the purpose of creating wealth for the individual and adding value to society".

British Pubs, "Where Friends Meet", are also Part of Britain's Culture

There are cultural similarities between people from all over the world, but at the same time, they can be vastly different.

Particular icons are easily identified with certain people and certain cultures. For example, the food icon can easily be associated with France, China, Thailand and Italy. But surprisingly enough, inasmuch as we know about English fish and chips wrapped in newspapers and pork or steak and kidney pies, people tend not to associate British culture with food. On the other hand, no one could deny that British pubs are known throughout the world, not only as pubs where friends meet, but also as a part of British culture.

Culture is a strong factor that can evolve and unknowingly penetrate into people's minds and souls as well as influence a person's daily life. People in general are dearly in love with their culture and hold it close to their hearts; this is expressed through language, music, art, fashion, and even hair styles. In reality, any persistent behaviour or personal preferences can become part of the culture that governs our behaviour and makes us what we are.

195

However, bear in mind that although people share the same culture, they might not necessarily behave the same, because culture is only one of many motivational factors. There are others such as biological and physiological factors which govern our emotions, personal beliefs or disbeliefs, family, friends and all sorts of societal influences. In fact, even work pressures, excessive wants and needs can all have an impact on our behaviour. Equally so, even individuals brought up in the same family, and educated at the same school could think, behave and act differently – just because we are individuals. Collectively, culture can be a strong influential factor that prompts us to act and behave, at times explaining who we are and why we do what we do.

In many different ways, people in different parts of the world share the same culture. This is true both for the big ideas, like a love of peace, and helping others who cannot help themselves, to the mundane, such as enjoying certain comforts and particular conveniences. But unless they are flexible enough to see beyond their own culture and prejudices, people will always emphasize different behaviours leading to greater misunderstanding.

Once the author stayed at a Kyoto luxury hotel in Japan, and he found that the men's washroom in the lobby was equipped with two types of facilities. One was a sit-down Western-type toilet, and the other a squatting one of the Eastern type. A big Australian (he was on the same bus with the author) went into the cubicle equipped with the squatting one since the Western-type toilet was occupied. When he came out, his face was all red as if he was about to have a heart attack. He growled with a very strong Australian accent: "What a bloody 'culture'." Out of curiosity, the author went into the cubicle that he used, and noticed that the Australian gentleman had made a mess in and around the toilet. Judging by his expression, probably he had made a mess on himself as well. The story was not quite at an end. When the author mentioned it to a fellow tourist, the gentleman said: "Raymond, don't tell me; you might as well know that those who are used to the Eastern-style squat toilet will do the same 'squat routine' on the Western-style toilet and make a mess as well." This story and many others like it might sound funny at times, but they illustrate how troubles are caused where there are cultural differences, and people are unfamiliar and do not accept the difference. In a more serious vein, some years ago (it was sometime in the early 1990s), a US paper told a short story about a Japanese student who returned home after receiving a number of years of education in the US under the American system and school culture. Upon returning to Japan, he continued his studies in a Japanese school. One day his parents received a letter from the school asking them to tell their son not to ask so many questions in class, because his behaviour interrupted the class. Asking questions is an important part of the North American learning culture; in Japan students are required to listen to learn.

Throughout history, human beings have had many confrontations caused by cultural differences. In Canada, the infamous Quebec Bill 101, the "French only Language Act" (1981–82) passed by the Quebec National Assembly prohibiting people from using the English language for outdoor public displays, is a case in point. In fact, the massive Francophone Québecois movement to separate from the rest of Canada is mainly a consequence of their determination to make French culture a distinct society.

In the 1960s and 1970s, there was the "hippie culture"; this sub-culture, as people like to call it, attracted a lot of disenchanted youth who grew long hair, sang flower songs, did not work and lived on drugs, drifting and wandering. And as the wheel turned, from the 1960s, 1970's sub-culture became the elite "yuppie culture", a symbol portraying early success: MBAs, Acura cars, mutual funds, and stock market investments. Many people switched from one to another without blinking. For the past twenty years, another culture has been growing: "Making money through entrepreneurship." What a world! Just think, about some 200 years ago, the Scottish philosopher Adam Smith had to rack his brains to tell people about the "Wealth of Nations". It is based on making use of different combinations of land, labour and capital to create wealth and value, but it does not say anything about how to put money into the individual's pocket. Yet after we have watched the feature performance of the dance of dragons for only two decades, "entrepreneurship", a word some people could neither spell nor pronounce right, all of a sudden turned into a pot of gold for the money grabbers. So, just as well "entrepreneurship" is now being honoured as a culture in its own right, one match-made with money-making endeavours. Under the circumstances, and for the purpose of celebration, it is perhaps inappropriate to say: "What a world!", but "What a twist of gold!" Then, some people might say: "What's new, the Chinese have been saying all along for more than five thousand years: *Gong-Xi-Fa-Cai!*"

Business Culture

Culture, in the context of business, is a medium through which we can interact with each other based on a set of rules that have developed over time. Unlike value, culture is identifiable, visible, and reflected in the behaviour or actions of the individual. It is learnt from family, schools, work and society. Value, on the other hand, is the driving force prompting people to behave and/or act. This persistent behaviour and action, with a definite pattern, eventually evolves into culture. It is a natural process, and the reason why the media, and in particular television, through repetitive displays of violence (even when shown in the more philosophical context of karate or other martial arts) is a vehicle promoting the culture of violence. In another context, *Justice Bao*, a television programme from Singapore, portrays a different culture: justice prevails, evil acts will eventually be punished, even when perpetrated by royalty; this programme and others like it help people to escape momentarily from reality.

A corporate culture speaks what the corporation represents, with a sense of purpose that prompts corporate action. Therefore, unlike corporate value, a corporate culture is tangible; it can be touched, and transferred from the corporation to the individual. There can be many different corporate cultures, but if a corporation is a community of entrepreneurs, then it will have only one culture: the enterprising culture based on an idea that everyone in the corporation is a creative and innovative agent to create wealth for himself or herself, and add value (through corporate activities) to society.

The Cultural Differences between Traditional Corporate Managers and Entrepreneurial Managers

It is not a parallel comparison, like the story of the Australian fellow in a Japanese hotel, but there are also "cultural" differences in management between the traditional management culture and the entrepreneurial management culture. Over years of observation and working with other researchers, the findings shown in Table 10.1 are noted differences.

It is necessary to note that the cultural distinctions illustrated in Table 10.1 do not mean there is absolutely no entrepreneurial culture in corporations, or vice versa. On the other hand, some entrepreneurial cultures could be found in some corporate organizations, just as the traditional corporate culture practice could penetrate owner/entrepreneur managed firms.

Enterprising Culture

Professor Tan Teck Meng, Dean of Nanyang Business School, Singapore, gave me the responsibility of overseeing the publication of an international academic journal with a regional emphasis when I was visiting professor at the Nanyang Technological University. The first task for me was to find a suitable name to reflect the support and the development of an individual's enterprising spirit. I recommended that the new journal be named the *Journal of Enterprising Culture*; it was approved and implemented. Immediately after the release of the inaugural issue, I received a telephone call from a local newspaper; the caller claimed that she was an English language expert and watchdog for the incorrect use of the language. She told me flatly that the name of the new journal was grammatically incorrect. She said that "culture" is a plural form and "enterprising" is singular, and we cannot use culture for a singular. According to her, the name of the journal should be changed to the *Journal of Enterprise Culture*. To me, it was a surprise to hear someone actually saying the journal's name is inappropriate. I told her there was no mistake in using the word

Table 10.1 Cultural differences between traditional corporate managers and entrepreneurial managers

Attributes and related characteristics	Traditional corporate management view	Entrepreneurial management view
1. Perseverance	Followed Chinese General Sun Tzu's war strategies. The first strategy: when things get tough, let's get out fast.	Never quit five minutes before midnight.
2. Energy	Eight-hour work is normal, ten hours would be a drain, beyond that is too much; unless of course, you enjoy meetings, travelling, mixing pleasure with business.	Endless work, endless energy, much the same as having a newborn baby. You need to have super energy to take care of the child day and night.
3. Resourcefulness	Very resourceful, as long as every day is 4 July. Otherwise, look for a greener pasture.	Very resourceful, be it 4 July, or a black Friday. When you are up, you want to stay up; when you are down, you have nowhere else to go but up.
4. Confidence	If boss says so, we will go for it.	I am in charge, if I can't be, who could?
5. Ability to take calculated risks	No risk, no gain; no risk, no loss. Risk is to be avoided.	Risks are challenges, no risk, no gain; no risk, there is a cost, the opportunity cost.
6. Need to achieve	Parking lot, executive washroom key, stock bonus, handsome expense account.	To make the next million, be listed in the stock market.

Table 10.1 (cont'd)

Attributes and related characteristics	Traditional corporate management view	Entrepreneurial management view
7. Creativity	Creativity is good, but it has to be fitted into the big picture.	We live on creativity, it is a niche to substantiate our efforts to meet competition.
8. Flexibility	Rules are rules.	Good rules should always be flexible. We made the rule, we can change it, if change is necessary.
9. Positive response to chances	Chances are chances, much the same as skating on thin ice.	Life itself is taking chances. Getting married, having children, travelling abroad, of course, including skating on thin ice.
10. Independence	Independence has a nice ring to it, but how do I cope with internal auditors?	Independence means meeting both the top line (personal satisfaction, and the feeling of achieving what I want to achieve), and the bottomline of financial performance.
11. Foresight	Five years is long enough, but how will the investor see what I visualize in five years and beyond? Will I be able to reach the promised land, if I am dead?	It takes much longer than five years for my children to grow up and mature; the same is true for business. The challenge is how to balance five years within and beyond.

Table 10.1 (cont'd)

Attributes and related characteristics	Traditional corporate management view	Entrepreneurial management view
12. Ability to get along with people	Of course, we care about you; you are one of us. But some of these others ...	Our product is XYZ, our strength is people.
13. Response to suggestions	It is a good suggestion, but ...?	It is a good suggestion, tell me what we can do to make it happen.
14. Profit orientation	Profit is why we are in business.	I would rather not make the dollars if we do not believe in or could not support the product.
15. Optimism	Optimism is conditioned by how well informed we are.	There is always an element of uninformed optimism. If we are totally informed, we probably would not want to be in business in the first place.
16. Initiative	So many initiatives, so little resources. If an initiative does not meet ROI requirements, that initiative is not worth the effort.	All initiatives should be considered, and worth giving a chance. Failure is only a word, until the initiative is completely tested.

Table 10.1 (cont'd)

Attributes and related characteristics	Traditional corporate management view	Entrepreneurial management view
17. Problem-solving ability	We can never solve all our problems. One goes, another will come. In fact, the challenge to managers is nothing but solving problems.	In every problem situation, there are perceived opportunities.
18. Imagination	Imagination is too subjective, we need evidence.	Imagination is a source of creativity. Without imagination, the earth is still flat. The moon is still dominated by the moon goddess.
19. Strong belief in control of your destiny	The future of the corporation is in the hands of investors.	The future of my business is in my hands. I believe that I can make it.
20. Total commitment	Commitment to the board and our investors.	Commitment to myself. A self-imposed commitment, so to speak.

Table 10.1   (cont'd)

Attributes and related characteristics	Traditional corporate management view	Entrepreneurial management view
21. Low need for status and power	It is a symbol of success, if my office is located next to the boardroom and in the top floor of the building. Call me JB, TK or BR but better address me as Mr, Ms or Mrs. If I have a doctorate, make sure everyone knows me as a doctor, better still address me properly. Wherever I go, make sure I walk either in the lead or in the middle surrounded by other members of the "team".	When in need, I roll up my sleeves just like anyone else. Small office, bigger manufacturing (or operational) area means low overheads, high contribution.
22. Dealing with failure	Failure costs the company money, and erodes confidence in ourselves.	We do not purposely make mistakes or deliberately tolerate failure, but if we fail, it is an opportunity for us to learn from our failure.
23. Capacity to inspire	In our organization, we always have individuals who either do not motivate, or are unmotivated.	Motivation is inspiration, an effective way to inspire people is to be sincere, learn to share, and develop a passion for what people create.

Table 10.1 (cont'd)

Attributes and related characteristics	Traditional corporate management view	Entrepreneurial management view
24. Veridical awareness and a sense of humour	Business management is a serious under-taking, status has to be maintained, other-wise, subordinates will walk all over you.	Awareness is a sense of responsibility. Humour is the lighter side of life, it relaxes tension and improves harmonious relations between people.
25. Orientation to goals and opportunities	The corporate goal is to satisfy sharehold-ers' return on investment. Any opportunity not meeting ROI objectives is no opportu-nity to the corporation.	Goals are more than just making a profit. Opportunities will unlikely be knocking on the door for a second time.
26. Performance	Judged by superior.	Judgement in the marketplace.
27. Management style	Delegation: use of control system, proce-dures and manuals.	Hands on; personal attention and "home-made" procedures; personalized management.
28. Attitude towards making a decision	Sound out ideas to a few more people: bet-ter still, call a meeting, form a task force or a committee, and/or do a study.	Make the decision and take the consequences.
29. Decision-making behaviour	Seek out perfect information.	Try it out first and see what is necessary, then smooth out the rough spots.

Table 10.1 (cont'd)

Attributes and related characteristics	Traditional corporate management view	Entrepreneurial management view
30. Dealing with a crisis	Let's find out who is responsible: an investigator's approach.	Put out the fire first – a firefighter's approach.
31. Communication	Secretary screens incoming phone calls. Often only VIPs are put through. Many callers are asked to leave a message in the voice box, which may or may not be returned.	Oral, person to person. Answer every telephone call personally if possible; otherwise, take the number and call back later, normally within twenty-four hours.
32. Leadership	Position leadership: there is only one position, therefore, I am the only leader.	Task leadership: there are so many tasks in our undertakings, whoever has the ability to lead a team for the task, large or small, he or she is the leader. He or she will perform the leader's task, and make the leader's decision.

Sources: Attributes 1–15: Hornaday (1982); 16–19: Gibb, as quoted by Kao (1990); 20–24: Timmons (1977); 25–31: Kao (1989, pp. 11–12), 32: Kirby and Fan (1995, p. 257).

"culture" for a singular, as entrepreneurship and the enterprising spirit belong to the individual. For the purpose of carrying out the mandate of the journal, which was designed purposely to stimulate the individual's inner need to create and innovate, therefore it is really "enterprising" that fits the mandate and not "enterprise". She agreed without hesitation. This chapter deals exclusively with the individuals within the corporation, and it is each and every individual's enterprising spirit that makes it the enterprising culture of the corporation, and not the enterprise culture of the corporation. What is the corporation's enterprising culture? It is to make a corporation a community of entrepreneurs, with everyone in the corporation thinking, behaving, acting and making decisions as an entrepreneur. It should be observed that entrepreneurship in an academic context is a body or a set of knowledge, whereas enterprising culture is a subset of the entrepreneurship discipline.

The Creation of a Corporate Enterprising Culture

It cannot be disputed that value is part of a culture, but building a culture in a corporate environment differs from the development of a corporate value. In the case of value development, it is a collective wisdom developed from the bottom up, whereas the building of a corporate culture is the responsibility of the management.

The sub-cultures that developed and are rooted in the business entities

The creation of an enterprising culture is the responsibility of the corporate management; consequently, it is vital that corporate managers appreciate that within the corporate enterprising culture there are sub-cultures. In practice, some of these sub-cultures are, in fact, hindrances to the development of the enterprising spirit, and therefore should be changed or eliminated and replaced by other positive sub-cultures that reinforce the spirit of corporate enterprising. The following is a list of selected sub-cultures that could be found in corporations that limit or prevent corporate enterprising culture development:

1. The "but" sub-culture.
2. The "we can never make it" sub-culture.
3. The "don't trust them" sub-culture.
4. The "initiative is nothing, but trial and error" sub-culture.
5. The "passing the buck" sub-culture.
6. The rigid staff control sub-culture.
7. The no decision "decision" sub-culture.
8. The meetings, meetings and more meetings sub-culture.
9. The "grab and run" sub-culture.
10. The "we can't put the environment before profit" sub-culture.
11. The "telephone, the manager's daily bread" sub-culture.
12. The "I am the boss" sub-culture.
13. The ROI sub-culture.

All of the listed traditional corporate management sub-cultures are self-explanatory. In order to appreciate what impact these sub-cultures have on people, a few of the above are elaborated. These include the "but" sub-culture, the rigid sub-culture of staff control, the "grab and run" sub-culture and the "don't trust them" sub-culture.

1. **The "but" sub-culture**

    "But" is just a three-letter word in the English language, yet it is so powerful that it can destroy or amend a Constitution, overturn a government, change the course of history, and above all, turn a strong-minded entrepreneurial manager into an all-star corporate "yes" woman or man. "But" works in mysterious ways. Virtually every single person has been confronted with the "but" sub-culture. The following are a few illustrations:

    Situation 1: A proposal to change the packaging design after an extensive study found the product needed a new image to bring it back to the front of the product life cycle.

The "but" culture: It is a wonderful idea, it is exciting. *But* what happens if the public does not respond to the change, who bears the cost? Moreover, we may even push this product over the edge messing up our customers' loyalty to the product.

Situation 2: A proposal to change recyclable containers to reusable containers and promote a programme to encourage consumers to return the container for money.

The "but" culture: Wonderful, this is good for our "friendly environmental image". *But* how do we know whether customers will return the containers for money? Doing this will immediately increase our initial container cost, we have to change our entire system to accommodate the change. Moreover, all containers have to be recleaned; this costs money. What would the health authorities say about the reused containers? Then there is the problem with the Labelling Act, and among other things, we need warehousing space. These days, both money and space are at a premium.

Situation 3: Changing the name of a product.

The "but" culture: This new name is good, it gives the image that the product intends to project. *But* we have invested a lot of money and effort in the original name of this product; how can we possibly think of changing it?

Situation 4: To introduce a publication in the name of the corporation to promote academic research and encourage the development of young researchers.

The "but" culture: This is just what we want. *But* did you do a feasibility study? Who are the readers? Where do you get people to submit their work? How do you market the product? What is the estimated readership? Then, another fifty whats? Whys? Hows? ... Last, but not least, if we fail, we would be the laughing stock of the community.

Some time ago, there was an experiment (it was shown on TV) to show the audience the meaning of "frustration". The demonstrator placed a glass divider in the middle of a fish tank, then put a large fish on one side of the tank and a large number of very small fish on the other. When the large fish saw the small fish, it made every attempt to eat them, but failed; the reason was obvious. However, a few days later, the large fish appeared to have adjusted and knew there was no way to get to the small fish. The researcher then removed the divider so the small fish were swimming around the large fish, but the large fish showed no interest in eating the fish. Like the big fish, people are continually confronted with the glass wall of the corporate "but" sub-culture. How is it possible for anyone to take the initiative, when knowing so well, the best and only answer to action is "but"?

2. The rigid staff control sub-culture

As a corporation grows to a very large size, it adopts a greater degree of professionalized management. Hence, it needs specially trained staff to exercise rigid control measures that will ensure corporate performance meets its required, planned or budgeted objective, making financial control a top spot in the organization. The same staff control sub-culture applies to other staff functions, including legal, and personnel. It is not uncommon to hear from frustrated line or operational personnel angry expressions such as: "Who runs this bloody place?"

As staff members do not generate revenue from outside, their contribution lies essentially in:

1. How they see themselves in the corporation in providing needed support service for the line and operation personnel to better perform their respective tasks in the marketplace.

2. Assisting top managerial personnel to oversee subordinates' performance based on budget, plans, procedures, rules and by-laws of the corporation; in another word, control.

Strangely enough, as both support and control are important functions of the staff, staff members tend to view the control function as needing their attention more than the support function. The reasons may vary. One is that the supportive role does not give staff the satisfaction of "star performance" in the spotlight, whereas the control function, more often than not, provides them with some degree of satisfaction through superiority. As a matter of fact, in order to ensure the control function is effective, rules and procedures must be applied, to some extent giving very little flexibility. No flexibility implies no toleration of deviations and irregularities. In the eyes of the staff control, entrepreneurial activities are nothing but deviations and irregularities. Rigid staff control in any organization is a hindrance to entrepreneurial endeavours, and discourages enterprising activities. The simple reason is that entrepreneurship is simply not a part of the control game. On the other hand, if staff does not exercise their role to "control", what else would they do? To provide support to operational personnel is one of the staff functions, but then, how would it look in the eyes of the CEO, and the performance report?

3. The "grab and run" sub-culture

The "grab and run" sub-culture is an ethical issue. What it means is that there are high level managers in corporations who do not wish to give credit and recognition to important contributions, including innovative and creative ideas and undertakings made by lower level personnel. Worse still, they will take it (grab) and make it as their own (run with it). Because of their position of power, the "grab and run" is often unnoticed. If the "but" sub-culture is a hindrance to entrepreneurship, the "grab and run" sub-culture is no less than the condoned murder of entrepreneurship in broad daylight.

4. The "don't trust them" sub-culture

For whatever defensive or aggressive reasons, one of the common corporate sub-cultures is distrust of others. It is not unusual to see that in a corporate working environment, members of the big corporate family act and behave as if they must be always on guard, or to use an American expression, the practice of "cover your a ..." (defensive type management). "Don't trust them", "keep it a secret" or "this is only between you and me"; all of these have the effect of isolating people. The consequences are obvious; if people are not trusted in the organization, how is it possible to develop an enterprising culture in the organization, if "trust" is not part of the corporate culture?

The above list is by no means complete, there are other sub-cultures developed within a corporate structure. They work and become part of the system over a period of time. Unnoticed, they become part of the corporate culture, embedded in the system and difficult to change. Change is possible, but it would be the same as a person putting his foot in the wrong shoe; it needs two operations to make it right. The task for the corporate manager is therefore twofold, making every attempt to remove undesirable sub-cultures from the system and replacing them with sub-cultures that will support the corporate enterprising culture.

Sub-cultures that support the corporate enterprise culture

Like a coin, sub-cultures have two sides. The side, as described above, hinders the corporate enterprising culture. There is also the side that supports the enterprising culture. Very much like real life – there are bad guys but, thank goodness, we have a lot of good guys as well. What we need is to strengthen those that are good and, if possible, change those that are not as good, or even bad, into good ones. The supportive sub-cultures include:

1. The encouraging human initiative sub-culture.
2. The encouraging people to make discretionary decisions and support the decisions they make sub-culture.
3. The continuing search and pursuit of opportunities sub-culture.
4. The individuals are agents for change sub-culture.
5. The breaking of the repetitive roundabout cycle sub-culture.
6. The believing in people sub-culture.
7. The changing staff control to staff support mentality sub-culture.
8. The no "but", no "if" sub-culture.

The list is by no means exhaustive (see Figure 10.1). They are all important, although the no "but", no "if", the changing of staff control to staff support, and the encouragement and support of individuals to make discretionary decisions sub-cultures tend to hold key places in developing a corporate enterprising culture.

As illustrated in Figure 10.2, the creation process is based on the individual and the individual's personal value, reinforced by entrepreneurial attributes and skills that build the corporate culture.

The Spirit of Entrepreneurship in the Corporate Structure

Enterprising needs the drive of spirit. Enterprising does not mean that everyone in the corporation goes and starts up his or her own business; rather it means developing an enterprising spirit in every individual in the organization. Staff control hinders entrepreneurship. People need not necessarily work hard, but they need to work smart. Staff create jobs for themselves, first the controller, then, the personnel. In an entrepreneurial corporate environment, staff roles are important, but they also have the entrepreneurial role to play like everyone else. They should justify their positions

Figure 10.1   The development of a corporate enterprising culture

Sub-cultures to be removed	Sub-cultures to be strengthened	Corporate culture
The "but" sub-culture	No "but" no "if"	
Rigid staff control sub-culture	Encouraging human initiative	Corporate enterprising culture
No decision "decision" sub-culture	Encouraging people to make discretionary decisions	
and other undesirable sub-cultures	and others	

by placing much greater emphasis on their supportive role for development as functional experts rather than staying with the mentality of "everything and everyone should be controlled, as there is nothing else, but 'but' and 'control'".

Figure 10.2   Corporate philosophy, value and culture

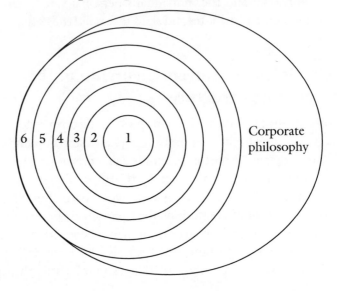

Legend:

1. Individual
2. Individual's personal value
3. Corporate value
4. Individual's entrepreneurial attributes
5. Individual's entrepreneurial skills
6. Corporate enterprising culture

11

Ownership, decision-making and corporate enterprising culture

Give a sufficiently long time, no one owns anything, give a suffi-
ciently short enough time, everyone owns something; that is, mak-
ing decisions and accessing resources under the individual's control.

God gives us a lifetime to access and allocate resources which we
do not own, so we must make good use of them for our own good, for
the common good of humanity and for all other living beings. It is
up to us to use and allocate the resources to achieve these objectives.
No less, no more.

Ownership, a Silly Little Word

Sometime during 1946 and 1947, the author had the rare privilege of representing the *Shanghai Evening News* in the press gal-
lery to witness Zhou En-lai (a moderate leader in the Chinese
Communist Party who once served as Director of the Political
Bureau of the Military High Command of the Chinese Nationalist Government in Chungking, during the Second World War)
and his team representing the Communist Party in China, negotiate with Zhang Qun (then Vice President of the Executive Yuan
of the Chinese Nationalist's Government) and his team representing the Government of the Republic of China. The negotiations
were about a possible solution to bring the country together and

215

rebuild the nation after more than six long years of Japanese occupation, while the memories and evidence of the ruthless killing and destruction were still fresh in people's minds. The negotiations failed, the subsequent civil war led the People's Liberation Army to drive the once powerful Kuomintang (KMT) and its government to Taiwan, an island province of China, separated from the mainland by the Taiwan Strait. In less than fifty years, Taiwan has shown impressive economic growth and claims to be the richest government in the world (but not necessarily with the richest people in the world). Meanwhile, on the other side of the Strait, China, the mother dragon, has made her economic recovery through the successful practice of a "market economy, with a Socialist governing system". Ironically, Taiwan is its second largest investor (in March 1996 Taiwanese investment in China amounted to US$200 billion). On both sides of the Taiwan Strait, the benefits of cooperation instead of confrontation, are contemplated. There were also signs of serious thoughts of unification through informal talks between business leaders of both sides. Unfortunately, it ended with Lee Teng-hui's private visit to the US in 1995, and his speech about Taiwan's intention of seeking greater international recognition, followed by China's military exercises in the Strait.

China's military exercises alerted Taiwan who responded by showing its own military capabilities. Their actions were similar to two brothers about to have a street fight, each one pointing at the other person's nose, just waiting for someone to deliver "the first punch".

Fortunately, no one took the first punch. With Lee Teng-hui's successful return to the presidential seat, there seems to be a good chance for renewed negotiations with China. Just like fifty years ago, there will be talks again. The only difference will be that this time the position of the players will have changed. Before, the Nationalists (KMT) were the strong men, but now the Goliath is the Communists. Will David use his sling to shoot down Goliath? No, not in this world, because the slingshot is no longer an effective military strategy; the new technology is human skills and the skills of negotiation. Taiwan has been proven to the world to be a smart businessman, as has China. And in business, everything is negotiable, the only difference is price. Is it all about market economy?

What is happening to China and Taiwan relations is, no doubt, a matter of ownership. While China considers Taiwan to be a lost province, Taiwan, at least from the public information expressed and interpreted by others, feels that it needs more international recognition; the bottomline is of course, "let me be myself". What is most interesting is where the US stands. It appears that its one China policy does not preclude support for Taiwan to be an independent state. On 20 March 1996, the US House of Representatives passed a "Taiwan Protection" bill which in effect placed Taiwan under US protection. All this is interesting enough as a topic of conversation, but the truth of the matter is still the question of ownership. Who owns Taiwan? (As the author sees it, to China, Taiwan is a province of China; to the Taiwanese, yes, Taiwan belongs to China, but they just do not want to be under the same government.) Who should make decisions for Taiwan? Perhaps even more so, who is making decisions for whom?

Nonetheless, this world is full of surprises; whoever would have dreamed in the 1940s that the US and Japan are now together strengthening their military cooperation efforts to cope with crises in East Asia in the light of tensions (March and April 1996) across the Taiwan Strait and the continuing instability in the Korean peninsula (*The Straits Times*, Singapore, 5 April 1996). Not too long ago, there was the West and former USSR Cold War, now it is between China and North Korea. Tensions or no tensions, the bottomline is still the "deep fear" of communism. On the other hand, if favourable environments are created so that everyone could have the opportunity to acquire ownership, who would want to be a communist anyway?

In the province of Quebec in Canada, René Lévesque quit his job with the CBC (Canadian Broadcasting Corporation) and formed his own political party, the Party Québecois, some 20 years ago. Since then, he and his followers have fought to make Quebec a separate nation; the struggle is all about ownership of land, of culture and of ideas. In Hong Kong, the people are struggling for the right to decide their own fate before the colonial era ends. There are more violent conflicts in the Middle East, Northern Ireland and Bosnia. Some of these conflicts are partially resolved: Yasser Arafat and the Palestinians in the Middle East, and Nelson Mandela and his people in South Africa have both succeeded in establishing at least some political autonomy after long battles, both political and physical.

By the time this book is published, all of the above will be outdated, perhaps becoming a part of forgotten history. But there will be new hot wars, cold wars and Yeltsin's "cold peace".[1] Judging from the way we repeat our histories, it is unlikely the struggles among ourselves will come to an end. And all of these are just because of one silly little nine-letter word: "ownership!" Yet, we know so well that given enough time, no one owns anything, because we will all one day be dead. But we do not know when that day will be. The Buddhists might say: "Poor soul who lived, and fought for ownership all his or her life, but when death called, it will be the end of it all. *E-Mi-Tuo-Fo!*"

The Buddhists may be right in saying that no one can take ownership into his or her grave, but at least one rich Japanese challenged this theory. Ryoei Saito stunned the world by buying a Van Gogh for a record US$82.5 million and a Renoir for US$78.1 million, along with other famous paintings in 1990. He hoped to be laid in a coffin with some of them after his death so they could be cremated together. He died of a stroke on 30 March 1996. Fortunately, his wish did not come through, because the "living" had no desire to be condemned by history for destroying priceless "human achievements". The poor soul was cremated alone. Amen! (*Newsweek*, 5 April 1996, p. 15.)

It is All About Making Decisions, Making Decisions for Ourselves and Affecting Others

Ownership is an expression of humanity's possessive, animal-like nature, luring human beings, individually or collectively with the power that "ownership" can give, but without realizing the responsibility that goes with it. For example, why should one person need 1,000 pairs of shoes? How many cows need to be slaughtered in order to make 1,000 pairs of shoes? Of course, the person may say, "I am not responsible for the killing of the cows. Where do you get hamburger meat, if the cows are not slaughtered? Leather is a by-product only." There are always excuses. Not just for the killing of cows, but also for causing harm to others. Yet behind the excuses lies the desire for ownership. Wars, dishonesty, cheating and other crimes are, almost without exception, rooted in the desire for more and greater ownership. It impairs the individuals' dignity, honour, and results in the loss of their sense of right and wrong. Just search our past, and observe what is happening around the world at present; were wars and human conflicts not prompted by people who desired ever greater and greater ownership?

There are many examples of individuals who desire to own everything, but seldom it seems, do these persons question themselves why? There are two simple reasons: one reason is to exercise the proprietary right to make decisions, to use them, direct them and allocate them in accordance with the rights of ownership. Ownership can also extend itself into shameful practices such as in the old days of owning human beings as slaves. Even today, people push the "possessive nature" of ownership to the extreme, always extending their perceived proprietary rights. They behave as if they own their wives, husbands, children and friends. Of course, inasmuch as the substance of owning is to make decisions and have access to resources, there is always the matter of formality and legitimacy. The other reason is recognition of ownership by others is part of the owning game – perhaps this is how we distinguish ourselves as civilized living creatures compared to other creatures, such as animals, birds, reptiles, fish and invertebrates with whom we share the natural environment. Could this be the reason why "civilized" human beings exercise their "perceived proprietary rights" by killing them and driving them out of their natural habitat?

Who has given us the right of ownership over other living creatures' right for life? Perhaps it is our "God given right". This is an argument which lies beyond logical reasons. If this is the case, though, it is our responsibility to decide how to use and allocate these resources – whether we make use of them for our own good, and the common good of humanity and all other living beings.

Corporate Ownership and Decision-making

In the traditional view, ownership is acquired by means of purchase, force, or granted by the law or through public recognition. The nature of corporate ownership has received considerable interest, from academics, investors, corporate managers, the stock exchanges, accountants, lawyers and, in fact, almost everyone in the community. Why is this so? The answer is obvious; corporations' well-being is the most important part of a nation's economy. Thus it is vital to know who owns a corporation, and how much of a corporation one owns (for example, stock exchanges want a full disclosure of the percentage of ownership held by specific group(s) or individual(s)).

Who has ownership over the moon?

When the astronaut Neil Armstrong planted the US flag on the moon, did this mean he had established in the name of the United States of America ownership over the moon? What if the moon were a liveable planet, and transportation from earth was available and affordable. Whoever landed on the moon first would most likely claim ownership before the dust settled from their first footprints. With no other contenders, the US would have ownership, and US laws would be applicable because it spent a large sum of money on its moon landing programme, and there are no natives on the moon to oppose them. All very simple, unless the ownership is challenged by another national flag. There are two possible scenarios:

1. A peaceful negotiation, determining who gets what. This could also be a possible solution, if there are natives on the moon (provided they understand human language); or

2. Two parties get into an insolvable situation, decide to settle this dispute with a fight. In the latter case, no one knows what will

happen at this point; they could destroy each other on the moon, even destroy the moon altogether, which is no good to anybody; or bring a nuclear war to earth.

Assuming nothing happens, then, a peaceful settlement of some form would have to be reached. But would we live happily ever after? Let's pray this would be the case. Oh! there is the "privatization" to be considered.

As everyone peacefully settled on the moon, the authorities decide to offer the moon for sale so the moon can become part of the market economy. Consequently, the government body involved in the moon project decided to incorporate under a state law (if the moon has no legislation that governs incorporation, may be in a state where MEL [Moon–Earth Link] spacecraft were launched, say, Texas) under the brand-new name of Moon Inc. listed in the New York, London, Toronto, Singapore, Moscow and Shanghai stock exchanges. Let us assume further, thanks to the blessing and magic of the market economy, our sub-planet moon governed by US laws and managed by Moon Inc. become prosperous and heavily populated. Other corporations are formed within the boundary of the moon. The moon economy continues to grow to be the one and only super "city" in the universe.

In the case of a corporation as illustrated (the Moon Inc.), on the basis of the traditional view, ownership is held by its shareholders[2] who either purchased, inherited, or were involved in founding the corporation. But while the traditional view may be suitable under the capitalistic idea, with so much concern about corporate power (like Moon Inc.), public intervention, consumers' concerns and environmental issues (assuming that the moon can no longer afford to absorb the growth pressure), we may wonder, what really is corporate ownership? How does corporate ownership affect our economy, society and livelihood in general?

The traditional view considers that only the shareholders own the corporation, but if a broader view is taken, shareholders are seen only as the "money" investors. Whilst shareholders invest money in the corporation, other people (employees) invest their life in the corporation, and the society invests their current and future stake in the corporation. A corporation is not made up just of "shareholders"; in fact, it is made up of a whole assortment of people with vital interests in it. Figure 11.1 is a simple illustration. The differences between the traditional model of economic entity and the idea of social entity approach are listed in Table 11.1.

Ownership is not an absolute, but a relative notion. For example, everyone breathes air. To breathe air is a decision (albeit an unconscious one) and action, therefore, in this sense we own the air around us. On the other hand, does it make any sense to claim that we own "air"? No one is able to claim ownership of air because there is a free supply. If anyone has absolute ownership, it is somebody superior than us, let us say: God. But God is not part of the market economy. On the other hand, if someone were able to capture all the air and store it in a tank, the free supply would stop. Through the market economy, we would have to pay for our air (just think of the person who had captured all the air, and the profit he or she would make). This certainly is not the case, at least for the time being. Remember, once upon a time, drinking water was clean and free, but some of us polluted the water. Today, if we want to drink clean water, we have to pay for it. Will there be a day when we will also have to pay to breathe the air we need to live? (Perhaps the government will like to nationalize it; if that is the case, how can the government nationalize Air Inc. if it is the only supplier of air? Because the one who controls Air Inc. can always cut off the air supply, then the government will be dead; thus, there is no way to nationalize air supply.)

Just as in the "air" story, there are actually two parts to ownership; one part is the statutory and/or public recognition of the traditional notion of ownership, and the other is ownership with the right to make decisions. In today's world, not everyone has statutory ownership, but everyone has the ownership right to make decisions either for him or herself, for others, or both. The idea of entrepreneurship is built on both the traditional statutory ownership and delegated ownership rights to make proprietary decisions.

Inasmuch as corporate ownership is based on traditional statutory and public recognition, it is expressed through the ownership rights to make decisions that matter to the individual, corporation

Figure 11.1 Corporate ownership

(a) Traditional economic and legal entity under micro theory*

People with money invest in the corporation* and acquire ownership
↓
and then become part or whole owners of the corporation
↓
Through shareholders' meetings delegate owners'
decision-making rights to the managers to make resource
allocation decisions on their behalf
↓
Managers accept the ownership's decision-making rights to make
resource allocation decisions.

(b) A social entity approach: A corporation is a social entity with
investment from the society**

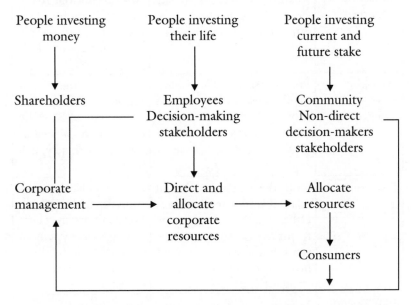

* A corporation is a legal entity; ownership belongs to the stockholders, therefore, making a corporation entrepreneurless.

** Under the social entity concept, a corporation is a community of entrepreneurs. The stakeholder is being perceived as having the proprietary right to make discretionary decisions when and where necessary.

Table 11.1 Corporate ownership: Traditional economic entity versus social entity approach

Area of difference	Traditional economic entity approach	Social entity approach
Ownership	Only those investing capital in the corporation	All individuals invest either their money, their life or stake
Decision-making	Decision rights delegated by shareholders	Decision rights delegated by all types of owners
Decision responsibility	Responsible to shareholders or those delegated by shareholders	Responsible to decision-makers themselves. Self-discipline, self-awareness
Allocating resources	Resources belong to the shareholders, hence, corporation through acquisition	Resources belong to the society, decision-makers accountable to society via senior lawyer decision-makers
Decision-making pattern	Primarily shareholders' interest. At times, managers must protect shareholders' interests at all costs	Interest for the corporation and those who invest and work for the corporation, as well as interest of society as a whole
Decision model	Only the managers should make discretionary decisions. Others may have the privilege to make discretionary decisions, but must follow rules, procedures and prescribed standards	All individuals have the right to make discretionary decisions. There is a support system to encourage individuals to make discretionary decisions
Staff function	More of exercising control	More support from all individuals to make decisions
Management control systems	Utilizing resources to achieve profit objective	Direct and allocate resources to create wealth and add value

and society. Therefore, the essence of the entrepreneurial approach to corporate management is to create an enterprising culture based on the corporate value consisting of three parts of managerial action:

1. To implement employees' corporate ownership participation, so members of the corporation can identify themselves with the corporation, not only as employees, but members (blood [ownership] related) of the corporate family as discussed in Chapter 10.

2. To delegate ownership decision-making rights to all members of the corporation, so each and every member can make discretionary decisions.

3. To focus on a corporate learning programme in conjunction with the stakeholder approach to facilitate employees' right to acquire corporate ownership rather than common shares. Details of the approach are given in Chapter 9 of Kao (1995).

On the basis of the above, the corporate culture within the scope of corporate value and philosophy may be expressed as shown in Figure 11.2.

Figure 11.2   A simplified model for corporate culture

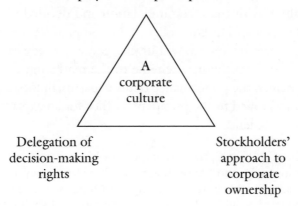

Employees share participation

A
corporate
culture

Delegation of
decision-making
rights

Stockholders'
approach to
corporate
ownership

The Topline Approach to Corporate Culture

Contrary to the bottomline approach to corporate management, the topline approach to corporate culture is a corporation with an enterprising culture in which every member of the organization will assume a proprietary position and make decisions no differently than any other owner of the business. The following are two illustrative situations.

Situation 1

I am making a proprietary decision, because I invested my life in this organization.

A large electronic company has seen rising sales in the past two years; but profit performance and the rate of return on the investment are not meeting budgetary objectives. A special team led by a management accountant studied the situation and diagnosed that the product warranty (five years for parts and labour) offered by the company had been so generous that it added an additional seven million dollars to the cost. Other manufacturers only offer three years' warranty, with the first year offering 100% coverage and the second and third year covering materials, but not labour. The recommendation to the company was to model their warranty on the competitors'. The team's recommendation was favourably received by the president, but the marketing manager refused to go along with the recommendation and decided to stay with the old warranty. The controller of the company let the marketing manager have his own way, but charged any excessive warranty service cost from the new scheme to the marketing department. Subsequently, the performance of his department looked so poor, that he was called to the president's office for an explanation. He told the president:

> I know, ROI and profit are important because we must satisfy our investors, and I know that we are here for money. But I am investing my life in this company. If I cannot stand behind our product, I would rather not have the money. I shall take this position and make the proprietary decision to continue with the warranty so long as I am the marketing manager. If this is not acceptable to you, then I have no alternative but to resign.

Situation 2

A foreign consultant had been brought in by an educational institution to help create a favourable environment which will encourage students to come up with ideas for new ventures. The proposal appeared acceptable to the administration. A meeting was called to discuss what to do to make it happen. At the meeting, a senior academic administrator indicated that he had this idea for sometime, but his plan was to encourage the development of ventures initiated by the faculty. Then, the discussion got into basic issues, and the administrator said to the consultant: "We have to look at the big picture, we must make a decision that has our country's interest in mind." The foreign consultant responded:

> Don't look at me as an outsider; as far as I am concerned, I am no different from you. You may be a citizen of this country, and I am just a consultant from the outside, but I am telling you, I consider myself having a stake here even if it is just for one day or one hour. My views and recommendations are the same as yours as a citizen of this country. Don't treat me as an outsider!

Stakeholder's Approach to Corporate Ownership

While the employees' share participation proposal permits employees to purchase the corporation's share stock at a price, in practice, most employees do not have the financial means to acquire corporate shares, even at a reduced market price. The stakeholder's approach to corporate ownership participation is based on the assumption that the employees' human efforts can be counted as value to be used to acquire a different form of corporate ownership, "stakeholder" as advocated by the author in his earlier publication (1995, pp. 159–62). Both the stakeholder's concept (in the illustration) and how to implement it may be summarized as follows.

Assume a large corporation has a paid-up capital of $800,000,000. The shares are held by various individuals and institutions, and they are publicly traded. The expected return on shareholders' investment is 20%, the annual dividend is about $10 per share, and the price per share with moderate fluctuation is about $100. Further assume the annual sales are about $4,000,000,000. Figure 11.3 shows a simple operating statement.

It is further assumed that the earnings and dividend payments are very much in line with the market trends. The $80,000,000 residual represents various individuals' contribution to the organization, not just the shareholders' capital stock at work. The no-capital contribution to the residual income is defined as the stake. (See Figure 11.4.)

To determine non-capital contribution in the stake, residual income is defined as:

Residual income = Accounting income − Capital charge

(The rate of capital charged must not be less than corporate bank borrowing rate. This ensures shareholders' opportunity cost is officially recognized.)

Figure 11.3   A corporation's operating statement (in million dollars)

ABC Corporation
Operating Statement
As of 31 March 2xxx

Sales	$ 3,000
Cost of sales	1,880
Gross margin	1,120
Less: commercial, administration and	
financial expenses	800
Net financial income before tax	320
Less: taxes	160
Net financial income	$ 160
Capital charge (paid out as dividends)	80
Residual	$ 80

Common shares fully paid = 8,000,000 shares at approximately $100
   per share = $800,000,000
ROI = $160,000/800,000 = 20%
Capital charge and paid out as dividends $10 per share = $10 × 8,000 = $80,000
Residual = Net financial income − Dividend = $800,000

Figure 11.4 An illustration of the equity section of the corporation's balance sheet before and after the implementation of the stakeholding plan

ABC Corporation
Equity Section
As of 31 December 2xxx

	Before	After
Common shares 8,000,000 @ $100	$800,000,000	$800,000,000
Operating residual	80,000,000	80,000,000
Total	880,000,000	880,000,000
Insurance of 80,000 stake shares @ $10		800,000
Total		$880,800,000

Equity share book value before issuing stake shares
= $880,800,000/8,080,000 = $109.01

Assume the corporation's decision-makers (most likely the Board of Directors) decide that employees' contribution should be recognized, but with a number of stipulations:

1. Of the residual, 10% should be allocated for employees' participation.

2. An employee is defined as a person in a contractual agreement with the Board of Directors, and not a member of the Board.

3. An allotment of shares of the stake will be made available for employee participation.

4. The price of the stake share will be determined by the Board, but it should not be less than 5%, and no more than 20% of the equity share price traded in the stock market.

5. Stake shares are entitled to the same dividends as determined by the Board.

6. Total amount of stake shares must not exceed 10% of the total equity shares traded in the market.

7. Stakeholders may elect one member to sit on the Board as a voting member (representing all stakeholders).

8. A stakeholder must surrender his stakeholdings to the corporation if he ceases to be a member of the corporation. The price of the share may be determined by the Board, but it should not be less than 10% of the equity share price traded in the stock market, or the price paid at the time of purchase of the stake price.

9. Stakeholders may attend the corporation's shareholders meeting, but they do not have voting privileges.

10. A stakeholder may not be a member of the union. In the event a stakeholder wishes to withdraw his participation, he may do so by surrendering his shares to the corporation at the original subscription price paid by the employee. He may then join the union.

11. The maximum number of shares purchased by any one individual may be limited by the Board.

12. Additional stake shares may be issued on an annual basis.

13. Stakeholdings may be redeemed by the corporation upon request from any participating employee, at a price of no less than 10% of equity share price traded in the stock market. If so desired, the redemption amount may be added to the employee's pension plan upon retirement, and the corporation may work out an annuity as it deemed appropriate for the individual.

14. The stake participation plan should not cause any great equity share price fluctuation. If, in any event, it affects the equity share price, the Board reserves the right to make necessary adjustments in order to maintain the company's share price at a reasonable level so it will not discourage investors' investment interest in the corporation.

15. Employees may elect a payroll deduction plan to purchase stake shares.

16. At the time of retirement, the stakeholder may elect to redeem the shares in cash (at no less than 10% the market value of equity shares traded in the stock market) or the shares may be left with the corporation on an annuity basis amortized until the retiree reaches the age of 91. Should the retiree die before age 91, the balance will be added to the estate.

17. Stake shares are non-transferable.

It should be noted that the issuance of stake shares is not the same as the employees' equity participating scheme. Both are meant to serve the same purpose: to provide an incentive for individuals in the firm to be able to identify themselves with the corporation, to be one of the three pillars to support a corporate enterprising culture.

The three pillars that support a corporate enterprising culture are: the employees' common share participation scheme, the delegation of ownership decision-making rights for others to make operational decisions, and the stake share participation plan. (See Table 11.2.)

In fact, any one of the three pillars can be found in many corporations, in particular the employees' common share participation. Combining all the schemes available will provide every member in the organization the right and privilege to participate in decision-making at all levels of the management process. This includes:

1. Making decisions as an owner of the corporation, through participation at the board meetings. The decisions are made without shareholders' prior approval, as the Board is only accountable during shareholders meetings. Although not everyone is able to take part in decision-making, views of the general populace in the corporations can be shared through representative participation. This form of decision-making participation gives a real accent of corporate democracy, even though on a representation basis. The management committee participation by employees (represented by members of the trade union) can be found in corporations based in the US and Canada.

2. Making decisions on the job can be troublesome, particularly discretionary decisions where there are no procedures or regulations to follow, or regulations which are too inflexible rendering operational processes difficult or impossible, or where changes need to be made based on the individual's personal judgement. The decision-making responsibility should be borne by the individual, but there has to be a managerial culture existing to encourage and support such decisions. The Board decision participating practice is now common in corporations based in Europe, and some in North America as well.

3. Making decisions as a stakeholder. As a larger representation may be required, stakeholders, through their representatives, can take part in both the Board of Directors meeting and the management committee. A management committee could be

Table 11.2 The employees' equity and stakeholding participation

Feature	Equity participation	Stakeholding participation
Equity position	Direct participation, from common shares. Allocation of common stock	Indirect participation, from residual. Allocation of residual
Rights	With all the common stockholder's rights	Right to receive dividend as determined by the Board of Directors. No general representation in the shareholders meeting; limited representation on the Board (only one), with no voting rights
Price	Based on common share price at a discount	As a fraction of the stock price
Relations with pension plan	No relations	Redemption value may be credited into the employee's pension plan
Transferability	Transferable, but with the corporation	Stakeholdings may not be transferred
Dividends and capital charge	Capital charge, the rate should not be less than corporate borrowing rate plus dividends	Only dividends
Redemption	Same as common shares	Only redeemable by the corporation
Union participation	May be a member of the union	May not be a member of the union, while holding stakeholder's shares

structured either as an advisory body or as a decision-making body. The participation of stakeholders in the management committee, not only gives the added decision-making strength, but a sense of belonging as well.

The challenge of the stakeholder concept is based on the fact that in a life situation no one owns anything forever. On the other hand, ownership exists as long as an individual is in possession of a resource, and is in a position to make decisions over the resources. In a corporate environment, every individual is the owner of his or her own doings or job, because he or she is in control of the job or situation.

Ownership is a relative notion; while a shareholder may own a fraction of or a whole corporation, and makes decisions about who should be responsible for the allocation of corporate resources, ownership will be terminated upon the person's death. Managers and all other employees are also owners of their own function, and ownership may end when the person's life ends. The same applies to individuals on the job; while on the job, ownership goes with the job and ends with job termination. Of course, if the worker stays for life, then the ownership will last just as long.

It must be emphasized that both in the democratic process and the process of entrepreneurship development, the key is ownership and decision-making. Decisions cannot be made without understanding the meaning of ownership; without the right to make decisions, a human being is no different from a working machine. The entrepreneurial approach to corporate management is the essence of an enterprising culture where everyone in the organization is a decision-maker and entrepreneur in his or her own right. A corporate enterprising culture is where people in the corporation have the right to make proprietary decisions for the purpose of creating wealth and adding value in the interest of the individual and for the common good. The aggregate value of all individuals and the corporate serving people philosophy is creative innovative wisdom. (See Figure 11.5.)

In summary, conceptually and in practice, a corporation under Entrepreneurism would have three types of ownership, which may be summarized as shown in Table 11.3.

Figure 11.5 From corporate philosophy to corporate enterprising culture

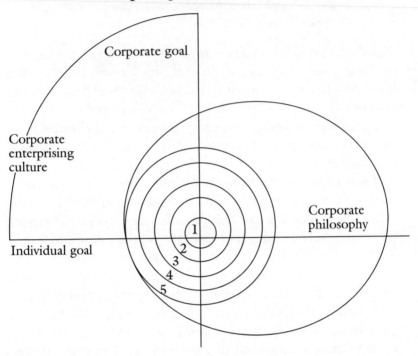

Legend:

1. Individual
2. Individual's personal value
3. Individual's entrepreneurial attributes
4. Individual's entrepreneurial skills
5. Corporate value

Table 11.3 Ownership under entrepreneurism

Type of ownership	Created/recognized through	Remark
Share participation	Acquired through the purchase of company shares (cash payment or through payroll deduction plan).	Normal ownership status, has voting power at the shareholders meeting. May be elected to the board.
Stakeholding	Received through allocation of the company's stakeholding scheme (redistribution of residual earned by individuals in the corporation).	Has an ownership status. It is an allocation of residual (after recognizing "full" cost of doing business, including capital charge). The scheme is a redistribution of the "fruit" of labour in addition to the cost entitlement of the labour, capital investment of shareholders.
Decision-making through delegation	Acquired through proprietary decision-making right through delegation (to view ownership is all about making decisions, once an individual is given the right to make decision, he/she in effect, has the right to make proprietary decisions).	Individuals perceive themselves to have the right to make decisions as they are proprietors of their jobs, tasks, assignments. They also assume the risk and responsibilities, accept the consequences associated with their decisions.

12

Mindset, entrepreneurial attributes and their development in a corporate environment

Entrepreneurship is all about people. It can't be developed through complaints by the CEO, or criticisms of a lack of drive, creativity or adaptability. It has nothing to do with pushing the "corporate vision" on unwilling employees. It is all about the challenge of encouraging commitment, getting individuals to say: "I (we) want to make it" that makes the corporation a community of entrepreneurs.

No One is an Entrepreneur All the Time, But Everyone has been an Entrepreneur at Some Time

It is a common notion, particularly among people in the academic community and business world, that entrepreneurship cannot be taught. And there are others who tend to separate it into the entrepreneurship mindset and entrepreneurial skills, by saying: "While mindset cannot be taught, entrepreneurial skills could." Some thirty years ago, like many others, the author accepted this, but he did so reluctantly because he was convinced that we are the product of our environment. If the environment can change, so can our

236

mindset. Throughout his lifetime, the author has seen many individuals suffering from poverty, with no drive, no energy, and no self-esteem – their attitudes were, to say the least, defeatist. Placed in an environment which encouraged and supported them, perhaps given a new chance through a windfall in the lottery, or a decent job, these same persons became go-getters, self-confident and, unfortunately, in some cases, arrogant. These are just personal experiences, but personal experiences which many people share. After exchanging views with friends at conferences, study sessions, and both primary and secondary research over a period of more than two decades, the author has realized that the most fundamental question is not whether entrepreneurship can or cannot be taught, or whether an individual can learn to become an entrepreneur. What we need to ask is what "entrepreneurship" really is, and who an entrepreneur is. If entrepreneurship is defined in relation to the process of creating a new venture, then, we must face the reality that not everyone is willing to start and manage his or her own business.[1] Moreover, does society need everybody to have a business of his or her own? Or is it even wise to encourage all young people to be owners/managers of their own venture? If the idea of entrepreneurship merely refers to "business ownership", it would be a simple matter to conclude that "entrepreneurship" does not exist in corporate structures, since only the stockholders are owners of the company.

From the corporate side, the challenge in a corporation surely is not to develop the "owning your business" brand of entrepreneurship. The development of spin-off companies is recognized as one of the options for individuals, either for personal reasons or corporate expansion, but this certainly does not mean that employees are encouraged to leave the corporation to start up their own businesses. In fact, if entrepreneurship is to be meaningful to humanity as well as to the individual, it has to be a matter of making a change, and an entrepreneur (as a Motorola's advertisement specifies) is an agent for change. Hayek (1937) regarded the entrepreneur as a person who responds to change, while Mises (1966) noted in his work that entrepreneurship is a characteristic not only of a business manager or a business owner. Even those who relate entrepreneurship with business undertakings have noted that only those who innovate and develop new combinations are entrepreneurs (Schumpeter, 1934). He pointed out that:

> Whatever the type, everyone is an entrepreneur only when he actually 'carries out new combinations' and loses that character as soon as he has built up his business, when he settles down to running it as other people run their business.

Looking at the matter more seriously, it would not be difficult to appreciate that since we can add nothing to our resources, nor to the environment which nature has provided for us, all human activities are a matter of gathering resources to make "different combinations", thus to improve their utility function for human use and other living beings. It is also a fact that no individual can be an entrepreneur all his or her life, but every individual can be an entrepreneur at some time. Consequently, an interpretation of entrepreneurship must not be confined to "business venture creation, and self-employment". Nor should only venture founders and business owners/managers be identified as entrepreneurs; all those who create wealth and add value must be included. For this reason, the author redefined entrepreneurship as being the wealth-creation and value-adding process in the interest of both the individual and the society (Kao, 1993). However, a definition should be both descriptive and prescriptive. Descriptive merely explains a fact, what is more important is its application: the ability to change people's mindset, and use it purposely.

Following the publication of " Defining Entrepreneurship: Past, Present and?" (Kao, 1993) and *Entrepreneurship: A Wealth-Creation and Value-Adding Process* (1995), a set of three tests were given to three separate groups of second- and third-year undergraduates at universities and colleges in Singapore, a completely unconstrained environment. The purpose of the tests was to determine whether people's mindset could be changed in respect to their understanding of "entrepreneurship" through a formal and informal learning process. One group consisted of approximately 240 second- and third-year undergraduate business and accountancy students; another smaller group of third-year applied economic students, and the third group was made up approximately of 250 students from both the business and technology faculties. There were only three questions:

1. In your own words, describe the term "entrepreneurship".
2. In your own words, explain who is an "entrepreneur".
3. Do you perceive that there is a difference between a course or subject in "Entrepreneurship" and "Small Business Management"? If the answer is "yes", explain why. If the answer is "no", explain why.

The results of the investigation at the time of the writing of this book were not yet final. The initial analysis indicated a change of mindset from believing entrepreneurship to be related to business applications to the broader definition that it should be applicable to all economic endeavours, including new venture creation and development.[2]

It must also be noted that a test using the same testing methods was given to a group of approximately sixty high school principals in Singapore. It showed a similar result; a high degree of shifting of belief in favour of a broad application of "entrepreneurship" was recorded in Wong Soke Yin's report in the *Nanyang School of Business News*, Nanyang Technological University, in September 1993. Moreover, the broad definition was also tested on corporate executives at various seminars given by the author during the past five years, with similar results.

All these investigations suggest conclusively that perhaps entrepreneurship cannot be taught, but by interacting with people and through formal or informal learning processes, an individual's mindset can be changed, thus appreciating the true meaning of entrepreneurship and its applications. Then, there is this question: What does all this have to do with a book on an entrepreneurial approach to corporate management?

Obviously, those who work in organizations need to change their mindset with respect to entrepreneurial attributes. These can be acquired though they are not familiar attributes such as self-confidence, perseverance and determination. A corporation cannot function in the marketplace with merely a CEO, directors of the board and/or with a handful of top executives. They need people, not one or two, but every individual working in the organization, if it is at all possible. Going back to our fish story: fresh, healthy water makes fish grow and glow, while stagnant water can do nothing more than kill the fish. Similarly, the employees' mindset must be in tune with the spirit of enterprising to make it possible for the corporation to move up to the frontier of growth and development. (See Figure 12.1.) More specifically, the processes critical to the mindset change in respect to entrepreneurial attributes include:[3]

1. Awareness that entrepreneurship is not merely about starting up a new venture or owning your own business, but is about being creative and making changes that will add wealth to the individual and value to society.

2. Realizing that we are energy, and energy is creative, therefore, we all are creative by nature.

3. Entrepreneurship is a way of life.

From Mindset to Entrepreneurial Attributes

If we compare the entrepreneurial mindset with the engine of a moving vehicle, then the entrepreneurial attributes are the necessary fuel to make the engine run. However, we must realize that all entrepreneurial attributes identified by researchers are not the exclusive property of the conventional connotation of business owner "entrepreneur", or of people in the process of creating their new ventures; all "entrepreneurial attributes" are also identifiable with others. For example, "independence" is often considered as an important attribute possessed by "owners/entrepreneurs". Would it also be true to say that every teenager wants to be left alone? Even a four- or five-year-old child is more often than not yelling at his parents: "Leave me alone!" Likewise, risk-taking, which is frequently cited as an entrepreneurial attribute that characterizes venture founders, is also applicable to other human beings. The only difference is that some have more and others less. It is also true that not everyone attempts to confront risky situations all the time, but everyone has had some risk-taking experiences. In fact, "life" is nothing but taking risks. Getting married is taking a risk. Even crossing a road at a pedestrian crossing is a risk, because a car

Figure 12.1　Entrepreneurship mindset and enterprising culture

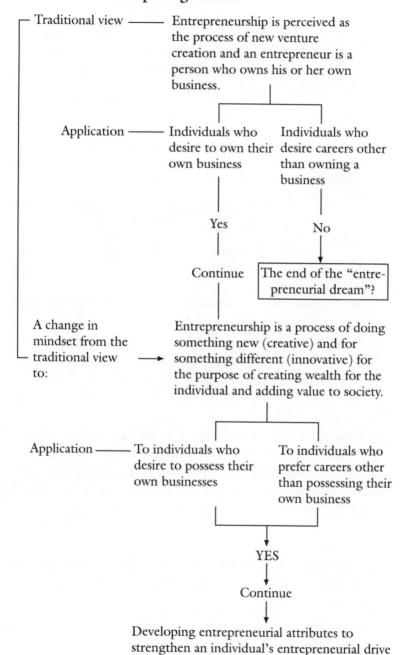

Figure 12.1 (cont'd)

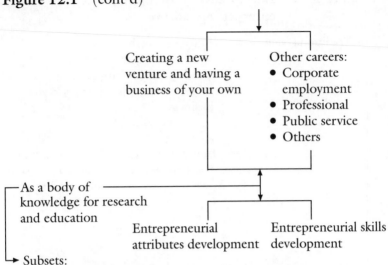

Creating a new
venture and having a
business of your own

Other careers:
- Corporate
  employment
- Professional
- Public service
- Others

As a body of
knowledge for research
and education

Entrepreneurial
attributes development

Entrepreneurial skills
development

Subsets:

- Family business
- Entrepreneurship for women
- Corporate entrepreneurship
- International entrepreneurship
- Parenting entrepreneurship
- Entrepreneurship for public service
- Entrepreneurial skills
- Others

or truck could still run you over. You may have just risked your life by "legally" crossing the street, yet the truck driver who almost killed you screamed at you: "Are you crazy? Just because you can be repaired, it doesn't mean you can walk right in front of my truck!"[4]

Entrepreneurial attributes can differ according to cultural values under different circumstances

Over the years, the author has made note of a large number of entrepreneurial attributes for teaching purposes and personal interest. He has found that all these attributes are everyone's inherent property, to some degree or another. Perhaps this much is obvious. A number of writers have noted how people from the East differ from those from the West. For example, risk-taking is a common attribute required for those venturing into the unknown, but individuals taking risks are often subject to their cultural influences. For people of Asia, anyone risking a venture into bankruptcy is unthinkable, but in North America bankruptcy can be a source of profit-making. And some Western ideas of entrepreneurial attributes are in fact, according to Kirby and Fan (1995), in conflict with Chinese cultural values. These include positive response to changes, taking initiative and personal responsibility, profit-orientation initiative, and a strong belief in the control of one's own life. Bear in mind, attributes are not simple matters that can be taken for granted just because "this is what researchers are saying". At least the following two points should be noted:

1. Some attributes are in fact, highly interdependent, for example, creativity is associated with initiative.

2. Some attributes lack congruence with each other. For example, it may be difficult to explain how a person can be both flexible and determined (Hornaday, 1982). The same goes for a low "need for status and power" vis-à-vis the "team builder and hero maker" (Timmons, 1977).

Attributes can be developed and should be developed, if they are essential ingredients that will enhance the individual's enterprising spirit needed to sustain the development of corporate wealth creation and value adding growth potentials. Then, we must ask ourselves what these attributes are, and how they can be developed through human efforts.

Developing Entrepreneurial Attributes as a Corporate Management Challenge

While this is a relatively new area for researchers, a few who have focused on investigations concerning entrepreneurial attributes include J. A. Hornaday (1982), A. A. Gibb (1986–87), J. A. Timmons (1977), and W. Gartner (1990, cited in Kirby and

Fan, 1995). Of course, there are others who have done the same although not all researchers provide a definite link between entrepreneurial attributes and success. Judging from the data collected and reported in the research works, attributes identifiable with entrepreneurs (venture founders and those possessing their own businesses) can also be traced to individuals who have nothing to do with business. For example, the "team builder and hero maker" attribute is also common to every social worker and "weekend party organizer". What matters under the circumstances is how these and other attributes in the individual can be cultivated, nurtured and developed into positive forces that will make the corporation a community of entrepreneurs.

So far no formal learning institutions have offered their services to develop entrepreneurial attributes for the individual. Although there are image consultants and industrial psychologists willing to lend a helping hand, the development process must be based on the willingness of corporate managers to create a positive environment, to encourage, provide support, take the initiative and make entrepreneurship everyone's habit to generate results. But bear in mind that with a few exceptions, most entrepreneurial attributes cannot be measured quantitatively: How much is "confidence" worth? What is the cost to humanity if all our "leaders" are malpractitioners in politics? Can we measure the damage done to the "goodwill" of France (which includes such seemingly little thing as two of my friends' refusal to go to the cafe' Delifrance for their coffee break), and to the environment because of France's political leader Jacques Chirac's "determination" and "perseverance" to go ahead with his nuclear bomb testing in the South Pacific? Could there be a nuclear war? If so, who is going to pin a medal on Mr Chirac's chest, if there are no survivors? (Chirac's call for the G-7 leaders to discuss and take collective action to deal with the unemployment problem was ignored by the other G-7 friends. Isn't that a strange coincidence?)

Nonetheless, attributes can always be put to good use to lead an organization to fruitful operation with a bountiful harvest both for the individuals and the common good as well. The challenge confronted by corporate managers is how to cultivate, develop, nurture and encourage these attributes to further corporate enterprising culture that supports our "social entities" to the benefit of us all. It goes without saying that shareholders' interests are definitely included.

What are those entrepreneurial attributes?

For the past two decades, researchers have been searching every nook and cranny to identify entrepreneurial attributes in order to support their claims of "what makes an entrepreneur". Kirby and Fan (1995) seem to have consolidated popularly recognized works. With permission, the summary is reproduced in Table 12.1.

Developing Entrepreneurial Attributes in a Corporate Environment

Not all 55 entrepreneurial attributes listed in Table 12.1 are relevant to the corporate environment, particularly those advocated by Gartner, even though according to Kirby and Fan, they are the

Table 12.1   Entrepreneurial attributes

Hornaday (1982) ■ 18 attributes	Timmons (1977) ■ 19 attributes
Self-confidence	Total commitment, determination and perseverance
Perseverance, determination	
Energy, diligence	Internal focus of control
Resourcefulness	Drive to achieve and grow
Ability to take calculated risks	Orientation to goals and opportunities
Need to achieve	Taking initiative and personal responsibilities
Creativity	Persistence in problem-solving
Initiative	Periodical awareness and a sense of humour
Flexibility	
Positive response to chances	Seeking and using feedback
Independence	Tolerance of ambiguity, stress and uncertainty

Table 12.1 (cont'd)

Hornaday (1982) ■ 18 attributes	Timmons (1977) ■ 19 attributes
Foresight Dynamism, leadership Ability to get along with people Responsiveness to suggestions and criticism Profit-orientation Perceptiveness Optimism	Calculated risk-taking and risk-sharing Low need for status and power Integrity and reliability Decisiveness, urgency and patience Dealing with failure Team builder and hero maker High energy, health and emotional stability Creativity and innovativeness High intelligence and conceptual ability Vision and capacity to inspire
Gibb (1986/87) **■ 12 attributes**	**Gartner (1990) Highest-rated** **attributes ■ 6 attributes**
Initiative Strong persuasive power Moderate rather than high risk-taking ability Flexibility Creativity Independency/autonomy Problem-solving ability Need for achievement Imagination High belief in control of your own destiny Leadership Hard work	Creation of a new business New venture development Creation of a business that adds value Integrates opportunities with resources to create product or service Brings resources to bear on a perceived opportunity Defines a creative idea and adapts it to a market opportunity

Source: Kirby and Fan (1995).

highest rated attributes. We cannot imagine that corporate managers can be convinced to develop employees' new venture creation attributes and encourage them to start up their own new ventures even though it might add value, and a spin-off is a way to claim that entrepreneurship can also flourish in a corporate environment.

Inasmuch as Kirby and Fan consider those advocated by Gartner to be the highest rated attributes, they are not necessarily relevant to the corporate environment, even though the creation of a new venture adds value. Therefore, for illustration purposes, only those relevant to the corporate environment will be used. However, as noted earlier, all attributes tend to be interrelated. For example, to develop leadership attributes, it is necessary to develop the individual's confidence and confidence is built on the need to achieve. It is associated almost directly with optimism, energy, diligence, and at the same time, determination and flexibility. Truth and goodness are also important, and the leader must always have the attribute of getting along with people. Then, which is which, how is it to begin? Unfortunately, using the "chicken and egg" story to explain the situation will not help. Let us try to use another "story", and see if it will help our "which is which" situation. The story goes like this:[5]

> A large corporation engaged the service of an internationally well known consultant. After months of work, the consultant submitted his report to his client. The report contained only one line, accompanied by an invoice (which had more writing than the report), for a sum of $100,000.00. The client was furious, and asked him: "I am sorry, what is this? A line worth $100,000.00?" The consultant replied: "Yes, $100,000.00, not one cent less." The client asked: "Why?" The consultant replied: "I am fully aware of the fact that everyone can draw a line. To draw a line is worth perhaps $0.01, but to know when to draw and where to draw the line is worth $99,999.99."

Under the circumstances, rather than considering which attribute is more important, or which one should be ranked first, let us just borrow the wisdom of the story, and use it as it fits. It should be noted that there are many other matured disciplines dealing with the development of an individual's attributes, particularly, in the areas of human resources development, industrial psychology and organizational behaviour. For our purpose, only a few are relatively more meaningful to the development of an enterprising culture in a corporate environment. These consist of three sets of attributes on the basis of their closer relationship in the developmental process:

- The first set includes the need for achievement (Gibb), confidence (Hornaday), perseverance, determination and dealing with failure (Timmons).
- The second set deals with risk-taking (Gibb), confidence and dealing with failure (Timmons).
- The third set has its core in leadership (Gibb), foresight (Hornaday), seeking and using feedback (Timmons), vision and capacity to inspire (Timmons), and ability to get along with people (Hornaday).

The First Set – The Need for Achievement Relating to Confidence, Perseverance and Determination

What is the point of improving efficiency and increasing company profit if the result is the loss of jobs, including your own?

Here is a scenario. The fundamental question in this scenario is a question of purpose: what kind of achievement is it to improve efficiency and increase company profit if it kills your own job as a result?

There is a steel company which manufactures all kinds of steel products including the large steel sheets and piano wires. Most of its operations are computerized, except its system for inventory control. Usually, good materials are used for industrial products, while the low quality coils are separated based on their tensile strength and durability, to be used to make screws, nails and strings for music instruments. The division, known as Roger's Work (not a real name) handles the materials and the production process and claims to have been a good profit centre that contributes approximately 30% of the company's total profit. The division is headed by a general manager and an assistant general manager. Measured by division investment, over the past five years it reached the required goal of 20% of return on internal rate of investment. Despite the recent recession affecting steel industries around the world, the overall company remained profitable. However, the Board of Directors and CEO felt it was necessary, as a preventive measure, to put the company on alert of possible setbacks, and launched an overall efficiency improvement drive. In Roger's Work, the assistant general manager (AGM) was actually in charge of the operation, and as part of his strategic plan, he investigated his own operation. He discovered that in the process of separating high quality steel material (in rods) from lower quality coiled rods, misplaced stock and poor labelling done by simply dropping rods in the wrong stock piles resulted in high quality music wire and industrial materials used to produce screws and nails requiring only lower grade and low tensile strength material. Among other things, he managed to have the operation streamlined and computerized, and revamped the overhead stock pile operation by employing a lift truck to place rods on a decentralized hanging system pinned on the wall. Total revamping and relocating production facilities cost the company about $150,000. After improvements were made, Roger's Work's division performance improved, and overtime in the shop was reduced substantially. At the end of two years, the AGM received a personal letter to congratulate him for a job well done. The CEO and directors of the company all received extra bonuses. Six months later, the AGM was told to leave the company; because of the changes in the stock piling system and the re-organization resulting in overall operational efficiency, the AGM's position was no longer needed and the entire unit became redundant. At a small private party in his honour, the CEO shook the former AGM's hand, handed him a Rolex watch, praised him as a great achiever, and thanked him for his contribution to the company. The AGM was every inch an achiever; his efforts to achieve higher efficiency cost him his job.

The case described above is merely intended to illustrate a point. Undue waste of human and material resources must be prevented, but is there any way to reward an achiever rather than tell him (so to speak) that he had killed his own job in the interest of the company? The sad thing is that some top executives and directors only see and feel "money" or "profit" which blinds their eyes and erodes their sense of responsibility to other human beings.

The former AGM, who lost his position by his own doing for improving efficiency, said this to the author:

> If I did nothing to improve the situation, most likely I would have remained as AGM for at least five or even ten years, because the company would need me. Now I have done it. I have improved the efficiency, re-organized the division, made more divisional profit, and contributed more to the company, but I also improved my position: "Out of the door, get into the open job hunting market!" Sure, I was told by my General Manager that the company would give me a good reference, but where could I go at the age of 48?

Developing a healthy environment that encourages creativity and individual achievement

Some people might say that the above example is an isolated incident. But it is a scenario reflecting real life happenings all over the capitalist world (perhaps less so in some countries such as Japan and Switzerland). Too often, with the motto "we compete or die", redundant middle managers and other people are laid off, discarded like rubbish. All in the name of improved ROI, to be shown in a fancy corporate financial report engraved in gold! And after the Board meeting, the CEO will raise his glass and say: "Gentlemen (or, Ladies and Gentlemen), let us raise our glasses to this year's remarkable achievement of reduced cost and improved ROI for our shareholders, and to our never-ending 'financial success' in the future." There is no mention of human misery, just another glorious success for capitalism.

Right or wrong, at a world conference on entrepreneurship held in Shanghai, China, in December 1995, one American friend supported the corporate position for improving efficiency and letting go the people who made it happen. He said:

> That is what capitalism is; we all compete in the marketplace, because it is the only unbiased mechanism we have to deal with

economic realities. Only those who are good enough can survive and prosper. Today, in the US to be a millionaire is not such a great thing as 30 years ago. Now, we have twenty or perhaps 50 times more millionaires in the US than in the 1960s. What's wrong with that?

Nothing is wrong, except if this is a corporate culture, why should people want to "achieve" anything? What do we say then, if achievement just means money to the CEO and directors, and the unemployment line for middle managers and Joe, Mary, Steven and Jean? We criticize those who swing to the left to the socialist's ideology, where there is no incentive for hard work, intelligent work or smart work (you may get a medal, but not money). On the other hand, in order to make the capitalists happy, smart work, intelligent work and hard work means working yourself out of a job.

By nature, human beings have an inherent desire to achieve. A child wants to achieve by climbing high, and all sport activities are based on the individual's need to achieve. Without young athletes wanting to achieve the expression of highest physical excellence and skill attainment, the Olympic Games would not be possible. And in recent years, in order to compete in the world market, Taiwanese corporations jointly with their government entered a new era of achievement. Instead of labelling their products as "made in Taiwan", all Taiwan products are labelled as "inovalue". It is a new term symbolizing the achievement of innovation giving the highest value to consumers who use their products. Similarly, all scientific discoveries are the result of the desire for achievement. Of course, sometimes artificial means of motivation are used with an inducement similar to giving a monkey a banana or a fish to a dolphin to motivate them to perform in front of a human audience. They are all tactics used to stimulate the individual's creative desire for achievement. Governments around the world give us countless incentives, including tax holidays, direct grants in capitalist countries and medals in socialist states; these are all part of the game to induce people to achieve. Therefore, desire and wanting to achieve is part of the nature of all living creatures. When discussing the issue with a few friends about whether pigs have a desire for achievement, one of them asked: "You are not a pig, how do you know they have any desire for achievement?" I responded: "Yes; but you are not a pig either, how do you know pigs do not have any desire for achievement?"

The need for achievement lies in human nature, and to stimulate and develop it in individuals in a corporate environment is the twofold responsibility of its managers. First, provide opportunities for members of the organization to develop their needs for achievement and second, and of particular importance, make sure that any decision and action on the part of the management must not discourage individuals' initiative for achievement. The "letting go" of the AGM example is a typical case in point.

A "quality management" expert on one occasion talked with a group of top level managers about what to do with individuals in the organization who have no drive, no desire to achieve high levels of excellence, or no sense of the importance of product quality. Put more positively, it is a question of how to improve the environment that will help members in the organization to be more creative and innovative. The consultant quoted a series of remarks made by Japanese academics and business leaders that expressed their concerns for the situation, including:[6]

> ... despite having the highest costs in the world, the United States is the world's leading economic power because it was able to produce a steady stream of innovative products such as cars, aircraft, telephone and computers.
> To remain competitive, the 21st century business community will require a creative self-motivated workforce with an international perspective.
> **Keidanren chairman Shoichiro Toyoda**

> The principle of "hammering down that nail that sticks up" which was prevalent in both Japan's public and private sector, destroyed creativity.
> **Professor Haruo Shimada, Economist, Keio University**

> What is needed is system innovation in education and business corporations in order to produce more creativity.
> **A senior MITI (Ministry of International Trade and Industry) official in charge of an informal study group for improving the business environment to foster creativity in Japan**

After the consultant finished the quotes, one member asked him: "What shall we do?" He said:

> We always want to learn from the Japanese, particularly quality management. Why don't we form an informal committee like what MITI has done to study how to improve the business environment to develop product quality and foster creativity?

Another member protested: "Oh! no, not another committee!" The consultant then said:

> Let me tell you a story: Some time ago, a corporate executive asked a wise man: "Dear wise man, what can we do to improve efficiency by about 20%?" The wise man thought for a moment, then said to him very seriously: "Let me see now, each one of us have five fingers on each hand; you want to improve product efficiency by about 20% and you have approximately 500 people working in your plant. The solution is quite simple, if you could find 500 people with six fingers on each hand to replace your present staff, it would increase your efficiency by about 20%." The executive was puzzled, but he asked the wise man: "Sir, you are right, but where could I find 500 men, all with six fingers?" The wise man replied without hesitation: "That is your real problem."

The story about finding 500 six-fingered men to overcome the efficiency problem is not real, but the wisdom which arises from it is clear. The creation of a favourable environment for people to commit themselves to achievement is the relevant goal and solution. But how do we achieve the solution? Thankfully, there are some ideas which do not need a committee of 500 men with six fingers on each hand. These include:

- To have a supportive attitude; instead of saying "no" to an initiative, say "yes". Any initiative is worth a consideration. Even though nothing will fail, it is also true that nothing can succeed, if it is not given a chance.
- To provide achievable opportunities at the level of the individuals, inter-company athletic competition is a good example.
- To compensate the efforts of achievement, either by tangible or intangible means.

- To encourage the need to achieve as part of the individual.

- To provide supportive systems within the structure. Although there is no need to form a formal structure, it could be informal counselling services within the existing organizational structure where assistance could be provided if needed.

- To set an example, making the company an innovative and creative organization and achieving a level of excellence in serving people, earning the trust of consumers, employees and society in general.

Interrelationship between the need to achieve, confidence, perseverance and determination and dealing with failure

Virtually all entrepreneurial attributes are interrelated to some degree. Some of these interrelationships are illustrated in Figure 12.2, which is based on a variety of sources, including Hornaday, Gibb and Gartner, conference papers and discussions with interested groups, such as the ENDEC World Conferences and other international conferences on small business and entrepreneurship.

Although Figure 12.2 is a simple illustration, the interrelationship suggests a reinforcement from various attributes. Success in one breeds success in the others, and as the individual has had the taste of success, more needs to be achieved. Further success will lead to higher achievement. (See Figure 12.3.)

The second set, risk-taking, confidence, determination and dealing with failure

"No risk, no gain, no risk, no loss" – this was a "motto" that about two years ago some conservative business persons were so proud of when dealing with business expansion and competing on a global basis. The "motto" has at least two possible implications from the standpoint of the enterprising culture. The first is, it may be appropriate to say, no risk, no gain, but it is not necessarily true to say, no risk, no loss, because there is the loss of opportunities. Without the pursuit of opportunity, no business can continue even at the level of survival under today's business

Figure 12.2 Interrelationship between the need to achieve, confidence, perseverance, determination and dealing with failure

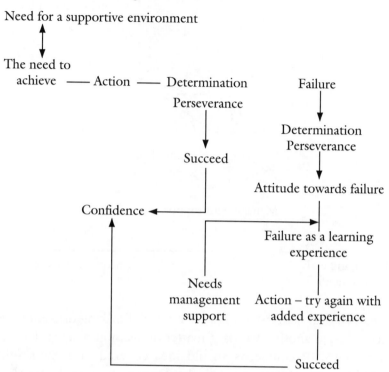

environment. The second implication is that while we all recognize that risk is an element that needs to be observed in any undertaking, it is the fear of risk that is often a greater problem than the risk itself. On the other hand, if an individual is in the midst of risk, and the desire and need is to deal with the risk, this makes risk no longer a consideration. For example, a non-swimmer may have a fear of water because of the fear of drowning, but once in the water, the need for survival will stimulate the individual to struggle to stay afloat, then risk is not a fact to consider. The second time around, the risk will not be a factor, whereas to learn how to swim is.

To overcome psychological barriers of risk-taking requires developing an individual's self-confidence. In a corporate environ-

Figure 12.3 Need to achieve illustration

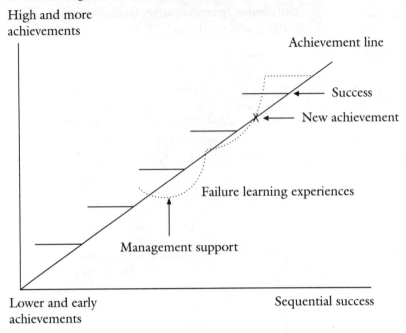

High and more
achievements

Achievement line

Success

New achievement

Failure learning experiences

Management support

Lower and early
achievements

Sequential success

ment, to deal with risk is a simple matter of making discretionary decisions, in another words, a matter of decision-making delegation. This, of course, is an old idea covered in every single management or human resource development book under the sun, but the strange thing is, it does not matter who said what, to let an individual make a discretionary decision is a thing of the past. Take the banking industry, for example. During the 1960s and 1970s, a bank teller could make a decision to honour a potential customer's cheque on the basis of his or her signature and satisfactory identification. Today, in Canada there are cases, even with a banker's (from another bank) cheque, and the bank's own customer, the teller acting on behalf of the bank must still hold the cheque and not release the funds until fifteen days after the presentation. The teller is no longer allowed to make that discretionary decision, the only thing that matters to the teller is to follow procedures. The crucial question to be addressed is that if an individual has no means to make discretionary decisions, how is it then possible to develop an individual's risk-taking attribute?

The bottomline solution is simple: with the infrastructure and corporate policy, allow and support all individuals to make discretionary decisions. Unfortunately, this does not often occur. In virtually all large corporations, the top level decision-makers rely heavily on procedures and systems to manage the firm. The thinking is even embedded in the language; a good employee is a "company man or a company woman". The reality is, if human beings are not to be encouraged and supported to make discretionary decisions, why then do we need humans working in corporations at all? When this was mentioned to a computer specialist, he said:

> Don't worry, computer software will take over everything under the sun. All you need to do is to follow the program. You don't need to think, or ask questions, because to ask questions, you have to read the program as well. In short, let the program do the work and make all the decisions.

In fact, we are already at this stage. When you make a phone call to an organization and try to locate anyone to respond to your inquiries or needs, you will hear: "This is J. J. Black corporation, if you wish to find ... please press 1 now; if you wish to speak to someone in marketing, please press 2; if you wish to inquire about our delivery service, please press 3." Then you press 3, and the sweet programmed voice says: "Thank you for calling the shipping department, our office hours are ... If you wish to trace a shipment, please press *; if you wish to speak to a shipper, please press 0." You press 0, then the sweet voice comes on again: "Please wait." You wait; the program provides you with soothing music. Five minutes later, "All our agents are busy, if you wish to ... please ..." The same routine can go on for hours, without the possibility of hearing a real human voice.

True enough, the computer does it all. Then, what about making discretionary decisions? There are no discretionary decisions to be made under the circumstances, including answering incoming telephone calls. Said the programmer: "Naturally, if there is no need to make any decision, where then is the risk?" But risk-taking is still a matter of reality, so, we continue.

Figure 12.4 From risk-taking to building up confidence

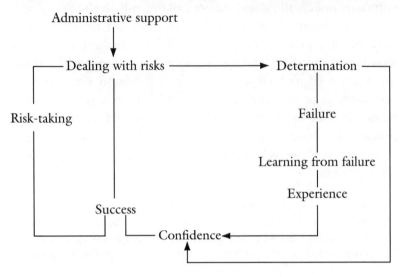

Figure 12.4 illustrates the interrelationship between risk-taking, determination, dealing with failure and confidence. In fact, virtually all entrepreneurial attributes have the characteristics of reinforcement, such as if an attempt is made to develop one attribute, it would affect the other. Typically, the success of one risk-taking endeavour would increase the level of an individual's self-confidence at the same time. Gaining confidence is a matter for celebration, but acting and behaving arrogantly would be deplorable. Although it is the responsibility of the corporate management to support an individual to further his or her self-confidence, it is for the individual to avoid arrogance while attaining a high level of self-confidence. One should learn to be graceful.

Leadership and other interrelated attributes

Leadership is not just a word. The human environment consists of leadership of different kinds, under different circumstances (see Figure 12.5).

As illustrated, the style of leadership changes depending on the leader's behaviour and decision-making. Autocratic leadership is

Figure 12.5 Leadership types move in accordance with circumstances

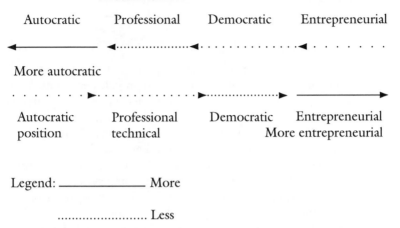

The illustration suggests that even the most autocratic leader could have, at times, some entrepreneurial spirit. Similarly, even the most entrepreneurial leader could be autocratic at times.

not autocratic all the time; similarly, an entrepreneurial leader can be just as "autocratic" as a six-star general in the front line of a strategic war. (In war-time we need five-star generals; in peace, we need six stars to deal with business.)

As communal creatures, we need leaders, and naturally, those who have the ability to lead rise to it when circumstances dictate. The birth of a leader and the subject of leadership can be seen in literature throughout history, but what is often not recognized is that the single most important challenge in leadership is leading without domineering. Even though this is a wisdom applicable to all leaders and in all leadership situations, it is even truer in the entrepreneurial environment.

Leadership is highly regarded by researchers as an important attribute, but the perception of leadership can be quite different, from the traditional point of view and from the point of view of an entrepreneurial organization.

The development of entrepreneurial leadership needs changes in attitude, with separation of rank or title from the task or mission, and recognizing that there are different types of leaders.

1. Leaders or leadership by type.

    There are many types of leadership, as well as many different types of leaders. There are people who provide democratic leadership (watch out, in a democratic leadership environment, it is not democratic all the time, as there is still the need for someone to make decisions) and autocratic leadership. Similarly, there are leaders leading people through their technical professional capability (most likely, autocratic leadership, because the situation requires technical or professional competence). There are other individuals who lead by virtue of their ability to adapt and react to situations immediately (likely democratic leadership, requiring individuals who are aware of people's needs and accepting others' views to make decisions) and take charge of the situation to lead people undertaking a particular task or function. Some individuals claim to be leaders either because of their positions, or because they are in an advantageous position which does not fit either the autocratic nor democratic category, such as a person with a gun in his or her hand. It is the gun (or a title) that gives the individual the command and domineering position, not the person under these circumstances.

2. A leader must have followers; a position leader is a position leader, not a leader who has a position.

    In any case, a leader can only be a leader if there are followers. Without followers there is no leader. In an entrepreneurial organization, a leader is not someone who stands on top of the mountain and tells everybody "I am the leader", nor someone accompanied by a team of staff, walking with them through the office or shop floor nodding his or her head and shaking hands with everybody along with a professional smile. Similarly, a corporate CEO, general manager or anyone in a high profile position would only be a "position leader" because he or she has a title, big office, and a parking space facing the main entrance of the office building. He or she may not have followers, hence, he or she may be a position leader, but not a leader who has a position.

    An entrepreneurial leader is someone who can always work with others and is a dynamic individual, with a positive, supportive attitude who believes that in every problem situation there are perceived opportunities.

3. Earned leadership and statutory leadership.

A position leader more often than not needs other forms of "statutory support" in order to attract followers, such as in the military environment, where the more gold bars, and stars, the better. In a corporate environment, there is written or unwritten code for executives; the name on the door, and the executive parking spot, all for the statutory image. However, there are smart executives who remove all the letters on the door, or simply place the sign "Private". In the academic world, of course, we have plenty of statutory symbols; the job title "professor" means a lot, the degrees and other honourable mentions, such as F.C.A. and F.R.S., all mean something of leadership to somebody who happens to be involved.

Can a person without any symbolic titles be a position leader? The answer is "yes". You would not see the president of a country giving a name card, or a "prime minister" carrying a plaque around his neck because he or she simply has no need for any such symbols.

4. The entrepreneurial leadership and entrepreneurial leaders.

Once a business friend asked me: "What is entrepreneurship; would a person who owns his or her own business be an entrepreneurial leader?" I answered:

> A business owner can be an entrepreneurial leader, if he has followers. When he speaks and acts, people in the organization would say, we listen to him, and do as he or she says, because he or she is right, and not because he or she is the boss, or at the least, we believe he or she is right and not because the boss has to be right.

This, in my observation is applicable to all leadership circumstances, not only for those who own their own businesses.

In an entrepreneurial corporation leadership exists within individuals, rank does not matter nor does position, title or whether his or her office or work station is in the office tower or in B6 (the parking garage). A fire warden is a leader once there is a fire in the building; a parking attendant is a leader regardless of rank or position. Everyone must follow his or her directions. A corporate manager is a leader, only if he or she is capable of directing and allocating and assisting people in managing resources for the common good including him or herself. How to develop the leadership attribute in a corporation is a simple matter of letting the "leader" have the freedom to make discretionary, professional, technical and other decisions and inducing followers as the result of his or her decisions relevant to the situation and the interest of the individuals in the corporation and the corporation itself. (See Figures 12.6 and 12.7.)

Figure 12.6   Leadership illustration

Position or autocratic leadership

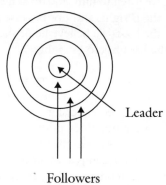

Democratic leadership

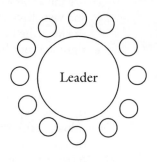

Professional or technical leadership

Followers

Entrepreneurial leadership

Each ◯ is a leader

Figure 12.7   Leadership and relationship with foresight, seeking and using feedback, vision and capacity to inspire and ability to get along with people

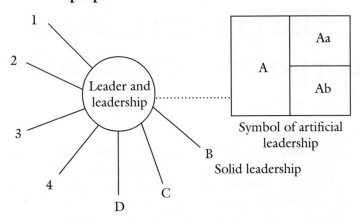

1.   Foresight
2.   Vision and capacity to inspire
3.   Seeking and using feedback
4.   Ability to get along with people
Aa.   Position leadership or
Ab.   Autocratic leadership
B.   Professional or technical leadership
C.   Democratic leadership
D.   Entrepreneurial leadership

13

The entrepreneurial corporate organization

An organization is a reflection of corporate culture, and an individual's identity. An entrepreneurial corporation is therefore a reflection of its enterprising culture and individuals can identify themselves with the corporation as entrepreneurs. Therefore, based on trust, an entrepreneurial corporate organization needs grassroot support and an effective networking environment among people that facilitates communication.

Organization Evolves with Time: We shall Begin with Ants

While we associate the act of organization with human institutions, it is not limited to them. The *New Oxford English Dictionary* (1986 edition) gives the following definition:

> The action of organizing, or condition of being organized, as a living being ... also, the way in which a living being is organized; the structure or organized body (animal or plant) ...

As such, we have the privilege of recognizing that ants are also organized living beings. Ants existed in their present form long before humans entered the world; probably, they will exist long after. And in all that time, while we may occasionally come across one or two ants on their own, usually, where you see one ant, there will usually be a multitude; tens, if not hundreds or thousands of ants, acting much like an infantry in military drill; orderly, effectively accomplishing one task – taking food back to the anthill. Ants have a history of organizational strategy which remains a mystery to us, and perhaps this will always be true. After all, we are not ants, and may never understand the ant's perspective.

Nevertheless, in some ways we admire ants, and we still study them and attempt to learn from them. From the human being's point of view, the ants have a simple organization, effective and meaningful. The effectiveness lies in its simplicity – a simplicity of purpose combined with a simplicity of function, served by a simplicity of organization.

Figure 13.1   Ants' organization

Ants (from various anthills)	Sources of food	In organized formation	Back to hills with food
Hill A ants			
Hill B ants			
Hill C ants			

By comparison, the organizational purpose of a corporation is much less clear. We marvel at ants, and wonder at the mysteries of nature which have created these incredible machines, yet at the same time we almost take for granted the organizational problems of the corporation. While all the ants want to do is to get sufficient food back to the anthill for the queen, the challenge to corporate management is to develop an organizational strategy, which considers human, material and monetary resources and reflects the beliefs of the managers themselves, and those who work with the managers. (See Figure 13.1.) The ants have it easy.

From Anthills to Business Organizations

They say there is nothing new under the sun. But if each life is not new, each single life, then why are we born? (Le Guin, 1974, p. 318.)

True enough, everything and every minute of life is new. On the other hand, if the question "why are we born" is addressed seriously, I doubt there is a satisfactory answer to the question except to those who believe in religion who might say: "We are here to do God's will." Unfortunately, such a "theory" can never be empirically tested. All we really know is that life is always changing, and even if the organizational theory has not changed, the organizations which the theory attempts to describe have. Perhaps the theory should catch up with the times.

Getting back to the ants; for the most part, we assume that the ants have not changed their hierarchy for millions of years: the queen ant is at the top of the pyramid with a variety of female workers below her, each worker accomplishing specific tasks. Human beings, by contrast, seem to change their organizational structure all the time. Although it is difficult for us to break the barriers built around ROI, some changes have occurred, particularly in recent years. For example, in this century there have been some significant structural changes within the capitalistic system. During the early and mid-twentieth century, corporations were (perhaps still are) owned by either well-to-do families and institutions or wealthy individuals. More recently, however, one of the single largest investors in major corporations are pension funds. In another words, today's capitalists are not necessarily wealthy groups or individuals, but organizations representing workers or retired workers who put some of their hard earned money away in "pension funds", which is now being channelled back to those large corporations. In some sense, we now have a situation where corporate management stands between the workers of the past and the workers of the present. Unfortunately, those making the decisions to invest the pension funds into the corporations are fund managers. They are not investors themselves which leaves corporate management still in the hands of block/majority shareholders or their appointees.

If we cannot make any real breakthroughs at the frontier of knowledge, at least we can improve or re-engineer the process of organization

How we can effect the changes to break through the management knowledge barriers or, for that matter, answering the question of why we are here, is really not an issue in this book. The issue and the purpose of this book is to explore how we can use the corporate experience to make life more meaningful for everyone – the corporation itself, ourselves and those around us. A true revolution is difficult and perhaps undesirable – it destroys before it can build. Evolution, however, implies meaningful re-organization to reflect corporate enterprising culture. As such, organizational theory must be part of the corporate management challenge.

From factory system to flexible organization and how business organizations have changed

The most significant turning point in the rise of the industrial era was the development of the factory system. The outstanding organizational feature of the factory system is an inflexible hierarchy, characterized by vertical integration and the separation of the individual workers from their entrepreneurial roots. But the factory system did not emerge fully formed into history. It is merely part of a process of continual evolution. While it is, of course, impossible to say where it all really began, the elements of the modern corporation can be seen in the guild system of the Middle Ages. Early attempts to solve the problems of cost, profit control, quality and output eventually made it give way to the system of entrepreneurial sub-contracting which was the immediate precursor to the factory system, which is still with us today in a recognizable form. However, just as earlier systems changed, the factory system has been superseded, first by the Matrix form of organization and more recently by the fashionable concept of the flexible firm. (See Table 13.1.)

Table 13.1 The progress of organizational forms (a historical view)

Organization	Feature	Purpose	Remark
Guild system	Entrepreneur-labour, capital-worker, internally homogeneous, hierarchically organized, low division of labour and simple technology.	Dealing with entrepreneur-labour, capital-worker relationship. Oriented to the end product.	No particular means of management, simple structure. Owner-managed business, not much delegation.
Entrepreneurial sub-contracting	Simple form of class with entrepreneur and worker, capital and labour, linked together in a "voluntary" fashion, distinctive from the guild system's caste-like nature. Home based.	Attempting to solve issues of cost, profit, control, quality and output.	As technological advancement makes it possible for individuals to start up their home businesses, the change of structure is likely to repeat the similar challenges confronted by entrepreneurs in the early days, only instead of owners of their businesses, now, they are corporate managers.
Bureaucratic pyramids	Factory system, characterized by vertical integration hierarchy. A boom of entity concept; within the entity, entrepreneurless.	Mass production and profit for the owners.	Until this day, we have not been able to depart from the system. Its organizational hierarchy is very much alive in virtually all corporate structures.

An Entrepreneurial Approach to Corporate Management

Table 13.1 (cont'd)

Organization	Feature	Purpose	Remark
Matrix	More flexible than the factory system. An organizational pattern involving multiple lines of decision-making in which both functional and project managers exercise authority over the same organizational endeavours.	Corporate profit, ROI for its stockholders.	There is no fundamental difference between a matrix organization and hierarchy organizational structure, except decision-making is decentralized to the operational level, thus facilitating further development into a flexible firm.
Flexible firm	Evolves with the market need. It claims that an organization should always evolve with a "rapture" that involves the "old" (Martin, 1974). A general concern of flexible firm organization is that these decentralized units are more often than not, without a single point of leadership; communication is horizontal, structures are cellular rather than pyramidal.	The same as above, corporate profit, satisfactory ROI for stockholders.	It leads to the development of management accounting approach to corporate organization, characterized as "profit centre".

Table 13.1 (cont'd)

Organization	Feature	Purpose	Remark
Profit centre	A concept developed through the application of management accounting, based on the assumption that a corporation is made up of a number of enterprise units. Each unit is a profit centre.	Division profit and the satisfaction of internal rate of return on investment, pressures for internal divisional competition for corporate funds. This leads to success in attaining corporate ROI objectives.	Led to the subsequent intrapreneurship development, including corporate entrepreneurship and spin-off, but no apparent difference from earlier systems, since the bottomline is still attaining corporate ROI objectives through internal accounting consolidation.
Entrepreneurial corporate organization	Recognizes that a corporation is a social entity, a community of entrepreneurs that every corporate member is able to make discretionary decisions, with a decision-making support system within the corporation. It features a difference in the fundamental corporate attitude towards people, resources, environment and profit.	Designed to facilitate innovation and creativity as a corporate strategy to create wealth and value for the individual, corporation and society.	Not a completely new idea, nor an organizational theory. Borrows from older organizational concepts, but has a more flexible approach to achieve corporate enterprising objectives with people in the same organization.

An Entrepreneurial Approach to Corporate Management

Are There Any Surprises in the Organizational Evolution?

As far as the author is concerned, there have been no real surprises in organizational evolution for over 100 years, and that is a big disappointment. In terms of the organizational hierarchy, it was more than 100 years ago that the pyramid organization structure of the factory system was developed and became established in England and a few countries in continental Europe, and possibly in other parts of the world as well. This system has been tinkered with, modified and studied endlessly; Frederick Taylor's system of scientific management, for example, and the development of Sloan-ism, which set the pattern of business organization from the days of Henry Ford, until the 1970s and the beginnings of the Matrix theory. Even this "improvement" has not changed the fundamental pyramidal form of business organizations, despite some significant modifications. In business organization, just as in management theory, we tend to follow the same course unless pushed, and in this case, the pyramid hierarchy remains king, and no one knows when we shall make a fundamental change, if indeed such a change ever proves necessary.

Of course, just as no two stories are exactly the same, so every twist to the organizational theory will show some differentiation, even if it is differentiation without merit. Does this mean that anyone who tries to say anything innovative about organizational theory in general will be guilty of over-generalizing? Certainly some people would say so. After all, they have the whole of history to choose from, and surely any critic worth his or her salt will be able to find one example to the contrary of any theory. It is easy to criticize, and hard to generalize without being at least a little wrong. Fortunately, in our "democratic" environment, we do not have to ask for forgiveness for being wrong. Just think how wrong our politicians often are; they generalize without apology, and usually without the merit of being innovative about it. Perhaps politicians should have a licence. What! Surely we do not need a minimum standard for politicians? Nevertheless, we will try to do what we can, not to be wrong.

The organizational casket

The pyramid organizational chart is found in every single management textbook and every single corporate annual report, almost without exception. As a matter of fact, all organizational charts box every single managerial and supervisory position of the corporation into a casket. All the caskets are the same size, perhaps with the exception of one or a few large ones to accommodate high-ranking individuals in the corporation such as the CEO and group general managers of some sort. It is the classical approach to all human organizations; we simply cannot get away from it. Then what is there to write about or discuss? Perhaps we have to begin with the beginning, and find out what is wrong with the boxed pyramid approach to corporate structure.

When the question was posed to a friend in business, he turned around and showed the author a piece of paper with one question on it: "Why did the chicken cross the road?" Then, he gave the author two answers, the first supposedly from Plato, and the second from Einstein. The first one said: "For the greater food"; to which he added: "For us, the greater the pyramid, the greater our profit." The second answer was: "Whether the chicken crossed the road or the road crossed the chicken depends upon your frame of reference."

Does the attainment of greater profit mean we need to have a bigger pyramid, or, if you have a bigger pyramid, will this get you all the profit (this is beginning to sound like the "chicken and egg" story)? Perhaps Einstein was right, and we need to decide if our frame of reference is the profit, or if it has become the pyramid? Or perhaps we need a new frame of reference altogether.

What is wrong with a pyramid organization that consists of a large number of small boxed individuals? The first thing is that the box only emphasizes communication with your nearest neighbours, especially in the vertical direction. Any communication within the boxed structure in theory would have to go through the line. (See Figure 13.2.) Just think what would happen to the bottom level manager who wishes to go through the organizational hierarchy with today's electronic "assisted" management system.

Note that if committees are part of the organizational structure, this may change its shape, but only to make communication

Figure 13.2   Going through the lines of communication

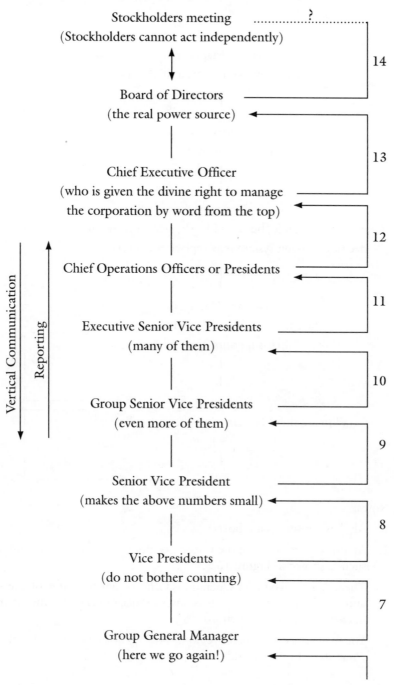

Figure 13.2 (cont'd)

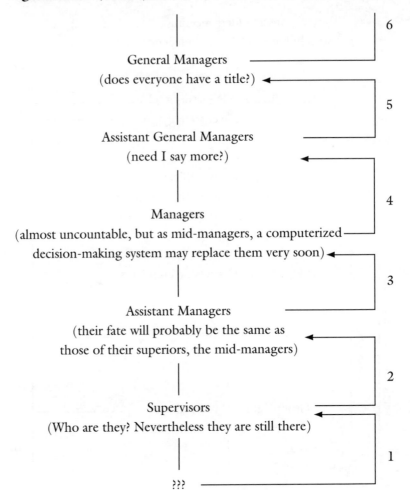

Notes:

1. All these positions are boxed.

2. At any level for each of the boxed positions can be a huge pyramid such as shown in Figure 13.3.

3. There can be two Chief Operations Officers, one in charge of operations including all product divisions, subsidiary operations, the other in charge of all staff functions.

4. You may notice that there are no workers. Who needs them? Let the computers do the "work"!

and decision-making even more complicated. Once Gulf Oil Company issued an organizational chart, for what purpose, I do not know – perhaps to demonstrate how "big" the company really is. It took almost a ping-pong table on which to spread it out. A big organizational chart like that would make a nice wall poster on the boardroom wall. (See Figure 13.3.)

Figure 13.3   Organization pyramid: Multiple of pyramids

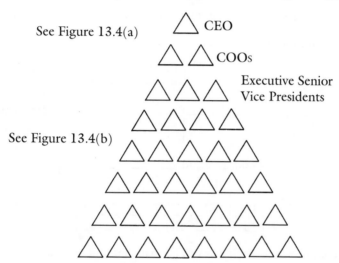

Each of the above small pyramids can be a huge organizational pyramid of its own as shown in Figure 13.4.

Organizational Hierarchy, Bureaucracy and Management Practice

If nothing else, the complicated pyramid organization has given "B" schools the opportunity to offer new courses, with the subject "business communication" during the past two decades becoming so fashionable.

With such a huge pyramid, helping people to communicate with one another becomes a huge problem; person-to-person contact is almost impossible. The solutions are many. Group meetings are one, and the telephone is often the most popular choice. Unfortunately this privileged form of communication is often kept for only a few key VIPs, with the rest of us having to deal with canned music and recorded directories. Then there are memos, both computer generated and e-mail. Even on the factory floor, it is possible to send a standard "personal" message to everyone on the shop floor. They are only workers after all; surely they do not need human contact? (See Figure 13.4.) To some people, to

Figure 13.4

(a) High level hierarchy in a corporate organization

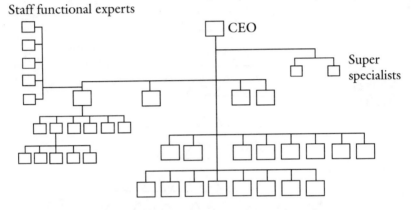

Boxes and boxes. Each box can be made into more boxes.

(b) Lower level corporate organizational pyramid

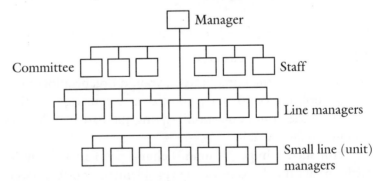

communicate or not to communicate or to communicate with computers or machines are not problems, but new business opportunities. It will not be long before the business communications department break away from "B" school to form a "School of Business Communications" by itself. What an opportunity this will be for business educators!

"Eliminate jobs" and "use more paper"; the twin princes of corporate bureaucracy

As the organizational pyramid gets bigger, communication becomes increasingly difficult. Person-to-person communication will become almost impossible at times and unless there is a good delegation system whereby individuals can assume proprietary responsibility, interpersonal relations would erode until a critical level is reached, where more harm rather than good will be done. This is illustrated in Figure 13.5.

Figure 13.5   Organizational pyramid and human relations

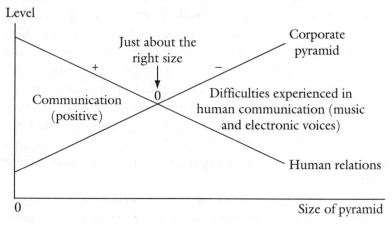

The straining of relations, increasingly tenuous lines of communication and ever more convoluted chain of command will make it necessary to call in functional expertise, to establish procedures and regulations to govern operations and/or using staff control to ensure the enforcement of procedures and regulations. This will give staff experts a sense of joy, as their responsibilities increase geometrically. Not only are they needed to prepare procedures and regulations, they are also required to interpret the violations and monitor the enforcement of rules and procedures. All that fancy management by objective stuff will be out of the window, and replaced with the management by fear along with management by crisis. Under these circumstances, once the individuals become indoctrinated into this system, I challenge you to find any motivation for creativity and innovation. Once this has happened, it will be easy to replace people with computers; after all, they would have been made into little more than computers themselves. So if your objective is to eliminate jobs wherever and whenever possible, to reduce labour costs and increase profit margin, then perhaps this is the way corporations should be managed. On the other hand, if you are in business that serves consumers, when people are without jobs, how is it possible for them to be consumers?

Then, there is the problem of paper, paper and more paper. Procedures and regulations need to be printed on paper; although this gives the duplicating machine company a market niche, what about the poor trees? And the garbage that wastepaper creates! The author once saw a poster in the Chinese countryside, which read: "For economic growth, it's us, for the environment, it's the government." For the author, concern for the environment is everyone's business. Government has a role to play, but we are all responsible for environmental deterioration. That includes every individual and every corporation, but it is particularly the large corporations who live off wastepaper because, without circulated procedures and inter-office communications, they would cease to exist.

Of course computerization has helped to eliminate certain secretarial and clerical work, but it has only exacerbated the wastepaper mentality. In some companies, ten e-mail memos are received for every memo people used to get. And people want hard copies (i.e. more paper!) of everything!

Last, but not least, it must be emphasized that procedures, regulations and rules are there to govern individual activities for conformity, and were never designed to accommodate creativity, innovation or human intelligence. Rules and procedures can never change by themselves; they need humans to initiate the changes for them.

Matrix, Management Accounting Model, and Organizational Apparatus under Intrapreneurship

The matrix

The matrix is a model designed to accommodate flexible requirements in dealing with project management and for those corporations who can divide their operations into enterprising units. It is certainly an improvement from the rigid pyramid hierarchy, where decision-making is to be based on lines of command. It nevertheless still subscribes to the same corporate management nightmare: to satisfy stockholders' ROI requirements, it pressurizes managers to make short-term decisions at the cost of the long-term wellbeing of both the corporation and the society.

The following is a sample matrix organization. What distinguishes it is how decisions are made within the corporation. In this case, the project manager and his or her team are decision-makers for the project; they assume responsibility as well. (See Figure 13.6.)

Figure 13.6 Matrix

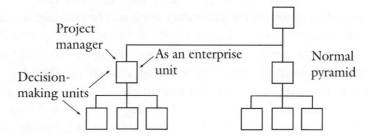

Management accounting (profit centre) organization model and intrapreneurship

Both the management accounting profit centre structure and intrapreneurship model place their emphasis on decentralized decision-making. They both are characterized by division autonomy within a given corporate structure, and without exception, both are subject to standard accounting and reporting processes. Both focus on division (or unit) profitability and internal rate of return on investment requirements. From the corporate point of view, both models rely on carefully engineered staff control. Hence, managers of profit centres or spin-off companies are still part of the pyramid organizational hierarchy. This is nevertheless an improvement, though perhaps only marginally. Re-engineering the process with an entrepreneurial emphasis can further the organizational change progress.

Whether it is possible to change the organizational focus from a "profit" "ROI" approach to corporate management to a more enterprising culture-oriented stakeholder organizational structure is a challenge to practitioners and academia alike, and not an easy one. Nevertheless it must be considered if the stimulation and development of corporate entrepreneurship in the business community is to be a fruitful reality.

Leadership and Corporate Organization

By and large, the nature of the organizational structure and reporting system tend to be correlated with the leadership style, though the connection is sometimes unclear. For example, an autocratic and/or position leader usually has a longer lifespan than, say an entrepreneurial leader, in the pyramid hierarchy environment. Autocratic or position leadership subsists on the idea of using staff control and (whether knowingly or unknowingly) the "management by fear" principle. They represent the status quo, and tend to make everyone do everything that maintains the status quo. This practice discourages new initiatives, and makes every new attempt fail before it even takes root.

On the other hand, a technical leader is more successful when the followers are all well-rounded, and motivated by special needs of the individuals. The democratic leader would be most suitable to a flexible organization, as a leader in a flexible organization leads others by working as a member of the team, rather than stay high above in the hierarchy. Entrepreneurial leaders tend to feel more comfortable and be more successful if there is a creative environment, and a great deal of interaction among people. (See Figure 13.7.)

Corporate Entrepreneurial Organization

In the human environment, a corporation is a reflection of corporate culture and the individual's identity. An entrepreneurial

Figure 13.7   Business organization variations and styles of leadership illustration

Autocratic and/or position leadership: Pyramid hierarchical organization (traditional)

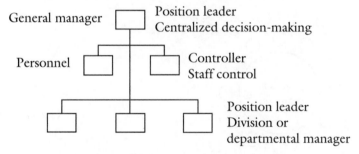

Circular organization and professional/technical leadership

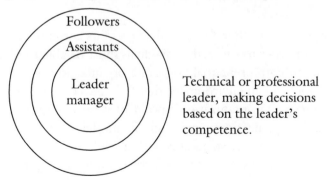

Figure 13.7 (cont'd)

Flexible organization and democratic leadership

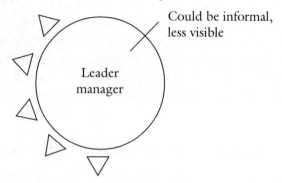

Spin-off organization and entrepreneurial leadership

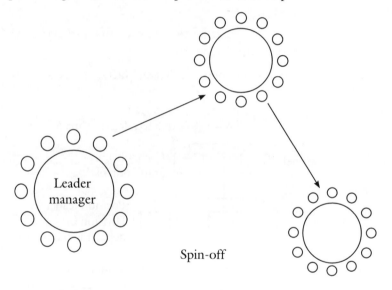

corporation is, therefore, a reflection of its entrepreneurial culture and, in order to succeed, individuals must be able to identify with it, as an entrepreneurial corporate organization needs grassroots support and an environment that facilitates communication.

From the organization's point of view, a corporate entrepreneurial organization is an integrated model consisting of both formal and informal structures with a triple mandate commitment built around a value-added centre as shown in Figure 13.8.

Figure 13.8 Corporate mandate in the corporate entrepreneurial environment

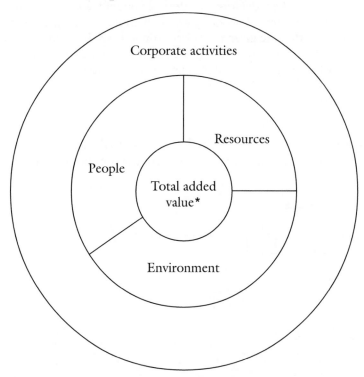

* Includes wealth to the individual, value added to society, and stockholders' interests.

Figure 13.9 shows a corporation as a social entity through its mandate of concern for the people, sensible allocation of resources and the protection of the environment. It creates wealth for the individual, generates return on the stockholders' investment and adds value to society. In addition, an entrepreneurial corporate organization is also characterized by the following:

1. A satellite formal and informal structure that embraces all four types of organizational structures: the traditional pyramid, democratic professional or technical and entrepreneurial.

2. Visibility: it is both visible and invisible. It has the inherited pyramid hierarchy with a formal structure with all the position leaders identified by their titles such as CEO, president, etc. For the

invisible part of the structure, it has smaller units with varied functional structures. When an organizational format is needed, it could be a democratic informal committee type of organization or a smaller working group, with no formal titles or ranks; if so, they are given on a direct voting basis. It could also have a structure for specific purposes, similar to a matrix organization, where individuals follow a leader with professional or technical competence. Leadership develops and evolves based on task needs.

Figure 13.9   Satellite formal and informal organization in a corporate entrepreneurial environment

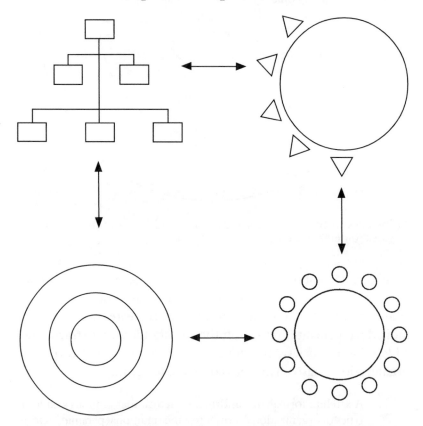

Note: All types of organizational structures could exist in a corporate entrepreneurial model based on need. They can be formally structured or informally arise as situations require, and they communicate through interpersonal relations both vertical and horizontal.

3. Human interaction: more human interaction – both vertical and horizontal. More often than not, offices and work stations are arranged in groups without walls (common receiving rooms exist when privacy is needed).

4. Decision-making: both centralized and decentralized. Discretion can be made at the point where things are happening, and decisions are needed. Along the hierarchy, a "line of command" remains as part of the organization, but is not necessarily visible. When people listen to the individual who makes the final decision it is because he or she is right, not because he or she is in the command position.

5. Measurement: individuals are measured by results, not the process. It could be both a bottomline and topline measurement. Whereas a bottomline measurement is based on a cost and benefit analysis, on the other hand, topline is based on the individual's inner satisfaction, whether a fully satisfactory bottomline solution is achieved or not.

6. Staff position: staff are identified as a supportive team. If expert authority is needed from the staff, the authority will be confined to a given task, and not beyond. The entrepreneurial corporation:

    - should act as a resources centre, providing expert assistance for all operational needs
    - should administer, minimizing paperwork for both themselves and others
    - should be a creative agent, continuing the use of a re-engineering approach to revamp or redesign corporate infrastructure to improve work efficiency, reduce waste, conserve energy and sensitive resources, minimize reporting waste, simplify accounting procedures and improve human relations.
    - must have staff that do not interfere with innovative and creative initiatives, unless significant corporate resources are involved in the undertaking, but are limited to assisting the initiative, helping to identify problems areas and provide needed support to solve the problem and/or identify new opportunities as a result.

"Trust", The Foundation of Corporate Organizational Entrepreneurial Strategy

Someone once said: "It matters not how much detail is embedded in a business contract. It is only as good as the people who stand behind the contract." Commitment and mandate are only words; behind those words are people, and it is people we have to learn to trust. It is the corporation in which we trust that a corporation becomes a social entity we can believe in and work with. This spirit of trust should be an integral part of all entities, both individuals and institutions, including corporations. A corporation must have the trust of its customers and employees, while management relies on the employees' trust to manage the firm. Of course, it goes without saying, that it is because of the shareholders' trust in the company's management that they invest their money in the corporation. It is also on this "trust" that a corporation formulates its organizational strategy. This strategy must build on developing employees as creative and innovative individuals. It must responsibly allocate resources with concern for the environment's health, while establishing the corporation as a social entity, accountable not only to shareholders, but to everyone that contributes to the existence and well-being of the corporation.

14

People, culture and skills

*In recent years, Japan has been experiencing the same harsh eco-
nomical realities as Canada, the US, and other Western nations.
In Japan, the large corporations deal with the economic downturn
by placing the importance of keeping jobs before profit. What about
the Western-based multinationals? Like General Electric's "Neutron Jack" who made his mark in history by killing 170,000 jobs
during his tenure as CEO of the company (Huey, 1992, p. 38).*

People, People and People

In the earlier chapters, we went to some lengths to deal with challenges relating to people in the context of entrepreneurship. These
include an individual's personal values, culture, the idea of stakeholding and participating in corporate ownership and, among other
things, the development of entrepreneurial attributes that could
make an individual more inclined to identify with and be part of
the corporate enterprising culture.

People as a subject are much discussed in the business and learning community. As an academic discipline, human relations and
human resources have been an indispensable part of Western learning and research at least since Abraham H. Maslow's *Hierarchy of
Needs* (1970). However, a Chinese scholar beat him by about 3,000
years. According to Confucius (not a theory as formal as Maslow's),
human nature consists of "food and sex" (in Maslow's terms, physiological needs and environmental are interpreted as air, water,
food, shelter, clothing and sleep). Nonetheless, the search for ways
to improve human relations and people management in industry

287

continues. Computers and applications of electronic technology have changed everything we know under the sun, but the issues and problems concerning "people" are still the most important, and are a long way from being solved.

From passion and caring to the wisdom of good deeds and self-discipline

While the subject has been touched upon in earlier chapters, a brief exploration will be made here of some of the very fundamental attributes which normally are not mentioned in others' writings. They are the attributes known as "passion" and "caring". Although these are common personal values that every human possesses, they are highly characteristic of good business entrepreneurs.

In one of my earlier publications, *Entrepreneurship and Enterprise Development*, Kelly Bulloch (the author's student) prepared a case for the book (1989, pp. 51–6). It was a personal account of her grandfather, the late John Bulloch, founder and owner of John Bulloch Tailors Limited, Canada. On his final day in his hospital bed, he requested a pad and left his last message. Scribbled on the pad was one sentence: "Tell the staff, their jobs are okay." The message was simple, and it made his passion and caring for his staff as clear as the blue sky.

If it is not too late to add a few words to Confucius' idea of food and sex as human nature, it is also passion and caring that are part of human nature. Although the simplicity of Confucius' wisdom of human nature is complicated by the new additions, they, nevertheless, are important for entrepreneurial attributes.

Most of us are familiar with the teachings of Confucius. Through his disciples, his influence is seen strongly in Japan,[1] North and South Korea, Taiwan, Singapore and Hong Kong, all of which share some common heritage and have in one way or another used and/or benefited from his philosophy in their business dealings, contributing to the overall nation's economic success. Application of much of his wisdom can be as beneficial to the individual. The author will list here a number of the more relevant ideas. Readers should bear in mind that though they are attributed to

Confucius, in some cases they are simply folk or societal wisdom, of untraceable origin. Indeed in some cases they are found in a number of different philosophies, both Eastern and Western. Whatever their origins, they are all worth repeating and considering:

1. Have a clear personal objective. If possible, ensure it is in congruence with the corporate purpose. If not, try to reconcile the difference and improve it.

2. Trust others, as you would have others trust you. An important challenge is that you must earn others' trust as others do not always trust automatically.

3. Be humble, but do not belittle yourself.

4. Have discipline, but be flexible when interacting with others.

5. To be powerful is not as good as to plan. Planning is essential to everything. Even though entrepreneurs characterize themselves as individuals with an element of uninformed optimism, it does not mean that there is no need to plan. A plan can be formal or informal; nonetheless, a formal plan is better than an informal plan which is better than no plan at all.

6. Give credit to those working with you, and should failures occur at least take a share of the responsibility.

7. Use others' strengths, and help others to overcome their deficiencies, or at least work around them.

8. Pursue your beliefs, and use whatever knowledge and skills to overcome the hurdles. Be focused and optimistic.

9. Care for those working with you, not just their work, health and personal problems but, among other things, their likes and dislikes. Do not impair their personal freedom, but act only if you are asked. It is the help that is needed, not the help you perceive they need.

10. Discipline yourself, and never raise your voice or show your anger to others who work with you. If there is a problem, the proper attitude is to consider it a mutual concern, and pull all resources at your disposal to resolve the problem, or better still, turn them into opportunities.

11. Search for views and opinions in respect to people and business. In most cases, these opinions and views are volunteered, honest and free. Accept them and make decisions based on their relevance and merits.

12. Make tasks easier for others, and yield to others for their convenience. Similarly, others are more likely to yield for your convenience.

13. Do whatever is possible to help others get out of difficult situations. Appreciate their disappointments and relieve their distress, if at all possible.

14. Work with nature, and to see nature's need as your own.

15. Perceive and believe others' problems to be your problems as well unless you are not aware of the problems, or others do not wish you to intervene with their problems.

16. Say less, but do more. Actions are always better than words.

17. Be decisive, but not headstrong.

18. Be a leader, rather than a follower. To follow the greatest is always to be the "second" greatest. Creativity and innovation mean to be ahead and not to duplicate others (exactly) or follow the yellow brick road.

19. Be confident, but not arrogant.

20. Protect others who work with you as you would protect yourself and your most precious personal values in life.

21. Respect others' responsibility.

22. Delegate work to those who work with you. However, delegation does not mean you do not care any more. It is only that you will have a greater responsibility for others to know what is to be done.

23. To criticize others aimlessly is a bad habit, in fact, even the "criticism" itself is of secondary importance. The need is to know about the situation to further understand the irregularities and reasons for the unpleasant happenings.

24. Appreciate the real nature of reward as well as discipline. To seek reward by generating short-term profit at the cost of corporate long-term benefit and the environment is a criminal act, even if there is no legal responsibility.

25. Have a good enterprising spirit, and continue the pursuit of opportunities.

26. Be a leader, and create an environment favourable to induce change, innovation, creativity, promote harmony, and improve good relations, thereby making the corporation not a money-making machine, but a community of entrepreneurs.

Developing people as creative innovative individuals

To develop people as creative innovative individuals both direct and indirect approaches can be contemplated. The more effective approach is to match individuals' interests and capabilities to their work, hence reducing the learning period and increasing the success in "work satisfaction". In a learning and informal leadership environment, individuals can commit themselves for "topline satisfaction" performance. One success could lead to further success, building confidence and the individual's inherent desire to break the barriers of learning, leading to creative and innovative undertakings. Figure 14.1 illustrates the linkages in our inherent creative nature through cultivating, matching, learning and providing developing opportunities.

As illustrated, the managers of the corporation have a clear supportive function in the above developing strategy. This includes both the line managers and staff experts. While line managers support the development process by providing needed encouragement, staff experts provide input (or conduct semi-formal seminars) and assistance to transform the strategy into workable programmes.

Organizational culture that affects people in an entrepreneurial corporation

People are products of the environment; we adjust our behavioural patterns to reach a certain level of comfort, establishing daily routines in life, and attitudes towards people that allow us to fit in. These patterns can include creative and innovative attributes. These are the necessary environmental factors that will facilitate the understanding, appreciation and development of corporate enterprising culture, with stakeholding and ownership participation as the motivational factor that drives an individual's desire for cultural and philosophical satisfaction through topline performance. (See Figure 14.2.)

Figure 14.1   Strategy of developing people as innovative creative individuals

Basic premise	Linkage	Method	Application
Creativity and innovation are inherent to human nature.	Cultivate, nurture and develop individuals' strengths and capabilities.	Observe and encourage human empathy in organization.	Provide opportunities for creative, informal leadership. Group activities in cultural interactions.
	Matching the work challenge with individuals' interests, preferences and capabilities.		
	Managers and individuals work together to explore individuals' need for creativity and innovation.		
	Learning process	Formal and informal	Learn individuals' likes and needs
	Provide opportunities for individuals who appear to have growth potential and are willing.	Based on individuals' commitment, determination and abilities.	On-the-job development, formal learning, informal person-to-person interaction.

S
T
A
F
F

A
S
S
I
S
T
A
N
C
E

Figure 14.2   Organizational culture that affects people in an entrepreneurial corporation

	Key ingredient	Function	The topline*
T H E	Managers that observe good deeds	Value-added focus Disciplined individual	Corporate enterprising culture
C U L T U R E	Corporate enterprising culture, understanding and drive	Motivation, individuals are motivated to create wealth and add value for both the individual and society	Management philosophy
	Stakeholding and ownership participation	Superior performance commitment	

*   The topline is a cultural performance that may or may not relate to bottomline (financial) performance.

Corporate enterprising culture, the marketplace and wealth to the individual and value to society

Some advertisers claim that consumers are kings and queens, meaning that the businesses are at the consumers' command. The intent is appealing, but the claim itself is deplorable; unless really, there is such a thing as "blue bloods", kings and queens are simply human after all.

A business corporation must serve consumers through selling products or services to their satisfaction. Currently, there is no visible evidence of any corporation making any effort to deal with consumers yet to be born. The strange thing is that we want children and we love our children, yet it seems that we do not care for them as we are not planning for their future. Perhaps it is even more striking that our plans for the future are not for the children's future sake, but to take what belongs to the future for our own present consumption.

An entrepreneurial corporation and its management must have in mind that its "serving people philosophy" includes serving the continuation of humanity. Our market economy must act to serve all consumers, not just those in the marketplace, but also our offspring, the human beings of the future. Figure 14.3 is an attempt to link the market economy of the present to the future and show how an entrepreneurial manager should perceive his or her role in building the linkage.

Corporate enterprising culture and management philosophy

Too often, the author has met people with the attitude that talking about people and the environment makes for an impressively philosophical discussion, but in the end it will be all about money. Money does everything; if you have money, you can even buy people. Quite a number of the author's students have had this view. The minute a topic on corporate philosophy, culture and value is mentioned, the response tends to be much the same: "Sir, you can say all these things because you have made it to the top, but what about us? Except money, is there anything for us to strive for? As you said yourself, we cannot get to the promised land if we are dead." The students do have a good point. "You cannot save a drowning child if you don't know how to swim." However, it should be noted that the students' view is based on their short-term need to survive, which is quite different from "greed". The challenge of balancing the long-term with the short-term is not strictly a financial issue, but a cultural and philosophical one with which we all have to live. To develop a culture that promotes entrepreneurship, meaning creating wealth and value for both the individual and society should not be a matter of choice, but the only intelligent and sensible cause.

Figure 14.4 shows how entrepreneurship provides bridges between people focused on management philosophy, value, and corporate culture leading to actions that build long-term relationships between the corporation and an individual.

Figure 14.3 Consumers are not kings and queens, but ordinary men and women just like you and me, our children, and our children's children

	Perception	Connection	Action
F A C I N G C O N S U M E R S	Consumers are not kings and queens, they are ordinary men and women. They have passion, and faults and are made of flesh and blood. There are also unborn consumers. Their blood is your blood, their flesh is your flesh, and everything of yours is in fact theirs. There are also silent business partners, the environment and the resources. They are fragile and limited in supply.	1. They have needs and wants. 2. They want people in business to satisfy their needs and wants. 3. It is up to corporate managers and others in business to decide how to satisfy their needs and wants. 4. To consider future consumers' needs and satisfaction the decision of corporate managers. 5. Suppliers in the market may take advantage of the scarcity of natural resources and human basic needs, charging high prices and making more than a reasonable profit, grabbing from nature to satisfy the present consumers' wants but not necessarily needs.	To sensibly allocate resources. To produce environmental-friendly products. To protect the environment from further deterioration. To balance long-term benefits with short-term cash requirements. To invest in creations and innovations. To develop corporate entrepreneurial spirit to create wealth and add value for the individual and common good. To be flexible, evolve with society and consumer's needs.

People, Culture and Skills   **295**

Figure 14.4 The corporate enterprising culture based on people focused on management philosophy

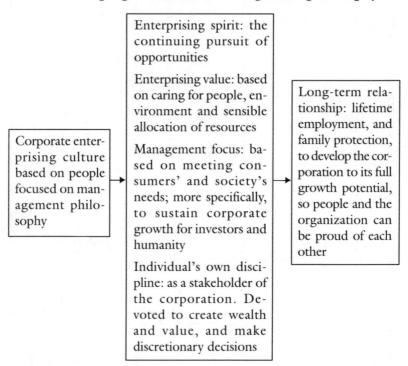

The Reward

In our zoos, circuses and research laboratories monkeys, lions, tigers, dolphins, birds, dogs and elephants perform, and after every performance are rewarded with their favourite foods; fish for the sea lion, bananas for the monkey, peanuts for the bird (maybe for the elephant too), and for the lion meat. On the other hand, poor performance is punished, such as in the case of the lions for whom there is always the lion tamer's whip. The system is simple enough and has analogies in human "management principles", performance rewards are based on the idea of "management by objective", and punishment like the whip correspond to "management by fear".

As humans, we have better reward systems than giving bananas or peanuts, but the principles remain the same. In fact, the entire classic economic doctrine is more a matter of reward (distribution) than production. Perhaps human conflict, including war, is caused by imperfect, inadequate and, mostly, unjust distribution. The rise of communism was caused by inequality and unjust "reward". Capitalism is also the work of a reward system. Without reward, why should anyone do anything? Just look around in all the business and management books; they all have done their share of covering the topic of rewards such as wages, incentives, bonuses, and stock options to executives. Under the circumstances, you can hardly avoid giving some thought to reward in our deliberations.

There are countless ways to reward good performance or attract loyal employees, though it is doubtful how effective some of the reward systems are in the long-term. The most common one is financial reward. In the socialist states and state-owned enterprises, good workers who performed well got medals and were treated like heroes. In capitalist society, other than money, reward can also come with titles. As a result, some organizations have as many as 50 vice presidents and directors, recalling the motto "if we cannot give you the money, you can always have a fancy title." You can only wonder how often the saying "We have no Indians, but all Chiefs" applies. Reward, be it peanuts, bananas, titles, money or medals all serve as incentives to motivate individuals to perform in accordance to expectations or better than expectations. Today, with some exceptions, money is still the one which does the most talking.

On 24 April 1996, Mr Wee Cho Yaw, CEO of the United Overseas Bank Group of Singapore, addressed the audience at the Universities Endowment Dinner organized by Nanyang Business School. Several parts of his speech directly addressed the matter of reward:[2]

> Take for example, the profit motive culture and its attendant notion that rewards must always be linked to profits. Since modern management stresses profit as the most important performance measurement, the staff is judged and rewarded according to the immediate profit they generate for the enterprise. This often leads to employees looking for short-term gains for the company without a thought as to whether their actions could lead to disastrous results in the longer term.

He continued:

> In the West, millions are being paid to professional CEOs to manage multinational corporations. They are well-educated and very experienced people. But again, since their benefits and their bonuses are mainly based on current profits, they would be more inclined to go for short-term gains by taking bigger positions even if this means taking on greater business risks.

He further emphasized:

> But I also worry that this cash nexus between employer and employee, resulting in hire-and-fire policies, could create more problems in the long term.
>
> My personal view is t hat sole reliance on cash incentives does not breed employee loyalty. In a tight market situation, we all end up paying more to the job hoppers, whose main aim is to earn the most money in the shortest possible time.
>
> I believe that, for their long-term interest, organizations should try to motivate and earn the loyalty of their staff by giving them a personal stake through share option schemes. For this reason the UOB Group was among the first to implement such a scheme.

What Mr Wee said is no surprise, but the message is clear enough: that times have changed and that short-term rewards may be good for monkeys, elephants, sea lions and birds, but perhaps not for humans. This is for the individual to decide: Should the company reward the individual with fish, bananas and peanuts, or with a long-term relationship (meaning lifetime employment) with the corporation? This would mean that the corporation would take responsibility to protect their employees (one of the good deeds), caring for their troubles, not only with regard to money, but also health, and family, among other things, making them real members of the entrepreneurial corporate family.

Last, but not least, there is the question of punishment (the whip). If a corporation is a community of entrepreneurs, motivated by a need for creation and innovation, making changes for creating wealth and adding value for themselves, the corporation and society, why is there a need for punishment? If punishment becomes necessary, why can't corporate managers try to practise another good deed by disciplining themselves first and be more understanding towards others?

Entrepreneurial Skills Development

Entrepreneurial skills

There are various skills appropriate to managers in an entrepreneurial corporate environment, including:

1. *Technical skills.* Skills about specific business, industry matters, the market environment, the competition and ability to deal with competition.

2. *Analytical skills.* Skills to assemble information, and derive a conclusion that leads to action.

3. *Conceptual skills.* Skills that lead to broad thinking and outlook on a macro basis. In other words, skills that give individuals broad abstract thinking, while keeping both feet on the ground.

4. *Interpersonal skills.* Skills to interact with people – networking. Such skills facilitate breaking the invisible barriers that exist among people.

5. *Communication skills.* Skills which include all forms of communication; written and oral communication, and even body language. Body language is in many ways the most subtle and effective, often a cause of misunderstanding, and leading to more harm than good.

6. *Negotiation skills.* Skills used in dealing with business contracts, and in situations involving matters with no obvious mutually acceptable agreement.

7. *Delegation skills.* More often than not, needed in any situation where tasks are beyond an individual's own physical and mental capacities.

8. *Developing relations skills.* Skills often expressed in difficult circumstances to bring people together, rather than to drive people apart.

9. *Skills development.* Skills needed to facilitate the development of skills in others.

10. *Leadership skills.* Skills that can often be witnessed in an unorganized group situation, where group members need

direction and help. An individual with leadership skills could quickly relate to other people and take charge of the situation. For corporate managers, leadership life often begins with a position and title, but position and title are not sufficient to make managers leaders. A few recent writers described leaders and managers as individuals who have more of certain personal traits or styles in relation to others. The role of a leader and manager can be a focal point in the organization for change, value creation, and, among other things, for advocating and promoting entrepreneurial activities that involve a stream of organizational dynamics and developing a human resources system (J. J. Kao, 1989, pp. 401–6). Some people recognize managers of this calibre as "social architects" (Kao, 1995, pp. 143–8).

Entrepreneurial skills development

Entrepreneurial skills can be learned from both informal and formal learning environments. Unlike mindset, skills development takes place in repetitive environments, and through interaction with other people. The same considerations apply to skills learned to meet work requirements or as personal assets.

For illustration purposes, six skills will be used, showing how can they be developed in the corporate environment. These include: communication skills, interpersonal skills, networking skills, negotiation skills, analytical skills and planning skills (Kao, 1995, pp.143–8).

Communication skills

Everyone can communicate, but communicating effectively, meaningfully and pleasantly is another matter. Most of our problems are caused by our inability to communicate. The US and the former Soviet Union never communicated effectively and meaningfully during the entire Cold War period. Effective and meaningful communication is the first step to peace on earth, happiness and prosperity for all. The following are communication skills that we should develop:

1. Be a good listener. Listening is communication. To listen is to allow maximum elbow room for the other person's freedom of expression. To have the desire and capacity to listen to others is an excellent communication skill. It is effective and meaningful, but how can you improve it? The answer is simple: be sincere, and listen sincerely as if you do not wish to miss a word. Listen to whatever the other person has to say even if the same thing has been said before. One frequent response from unconcerned listeners to the person repeating the story about his or her interesting adventures is: "Oh! You told that before." The fact is any person of sound mind would eventually realize this and stop the "torture" before going too far by saying: "Oh, I told you that before." If the torture is not too painful, encourage the story-teller to finish the story by saying: "Please go on, it's interesting." The exercise is a simple matter to allow others to feel free to express themselves.

2. Be caring, interested and responsive. As a listener, it is vital to impress the speaker with your reception of his or her words. Listening should come with favourable facial expressions, showing a genuine interest in what the speaker has to say.

3. Smile. The smile must be from the heart. Smile genuinely, not professionally.

There are also a few don'ts in communication skills development:

1. Avoid excessive body language. Body language often suggests displeasure towards the communicator. Even pleasant body language such as hugging, laughing and waving hands can often be wrongly perceived. Unfriendly body language can hinder effective communication, block communication and even terminate communication. Communication is not just what you say or do but how you are perceived.

2. Avoid sarcastic language. Statements such as: "I am sure you can do better than this", or "You have made enough mistakes for the day, you can go home now", should be avoided.

3. Avoid screaming or raising your voice. Screaming or raising your voice during normal conversation suggests that the communicator does not know how to communicate effectively using a normal tone of voice or a normal expression. This invites a screaming match, and can effectively terminate proper communication, inducing hostility.

4. Avoid interrupting the information flow. Always allow the other person to communicate and complete the delivery of his or her ideas, arguments and suggestions.

5. Avoid changing the topic. If it is necessary to change the topic, end the previous topic in a satisfactory manner and then introduce the new topic at the appropriate time.

6. Avoid any physical obstacles between communicators. Any obstacle can block the communication, and to be effective, if it is at all possible, communication should occur face to face.

Last, but not least, it is always desirable to communicate in a pleasant environment.

Interpersonal skills

Most of the communication skills and positive attitudes are applicable to interpersonal relations. An effective, good, sincere communicator with a positive attitude probably has good interpersonal skills. There are other areas of interpersonal skills development.

1. Learn how to give as well as how to receive. Some people only know how to give, and others how to receive. An entrepreneurial person should know both. Giving need not involve money or anything of material substance. Caring and sharing are all part of giving and receiving. The bottomline of how to give and receive is a matter of sincere appreciation and gratification of expression extended between individuals.

2. Approachability. Approachability is a simple gesture to make others feel that you are available if they are in need. "My door is always open" is good, but not enough; a more effective interpersonal skill requires a mutual feeling of "my office has no walls".

3. Building bridges not walls. The idea of building bridges and not walls suggests an individual should find a way to reach others or make it possible for them to do the same by building a bridge to allow others to cross. On the other hand, two individuals face to face can even have walls between them: a deliberate attempt to ignore the other's presence, harsh remarks, unfriendly body language and, among other things, refusing to communicate (such as not returning telephone calls) are cases in point.

Interpersonal skills can always be developed. The key to developing interpersonal skills is to interact with others on any occasion and in any place with sincerity and concern for others. Some people gain the skill in their childhood (through play and interaction with other children and adults), while others may not appreciate its need even when entering into adulthood, whether at work, at social gatherings, at church functions or in the marketplace.

Negotiation skills

In one airline's inflight magazine, an international consulting firm was attempting to make a point with a simple advertisement (featuring a photograph of a non-smiling person with a successful look and looking straight ahead) by saying: "In business, you don't get what you want, you get what you negotiate." The interesting thing is that many of these important skills are not part of the formal learning process; individuals have to learn them by themselves. There is another popular saying in business: "In business, everything is negotiable, the only difference is price." Is this true? No wonder Robert Redford's character in the movie *Indecent Proposal* thought that with a million dollars, he could negotiate for someone else's wife's companionship for one night.

Negotiation is a skill, but it is not a skill with the intention of making a killing. Good negotiation skills must be supported by sincerity, good faith and the intention that the end result will be fruitful to both parties. A sincere negotiation will not have one party beginning the negotiation from the floor (the bottom), and the other from the ceiling. It is rather a reasonable approach to assume the other's sincerity until proven otherwise. A friend of the author once told him to watch out when entering into negotiations with "some people" in Hong Kong. He said: "At the end of the negotiation make sure that they haven't taken your arm away with them." If this is true, it is certainly not an entrepreneurial type of negotiation for a person to take someone's arm away as part of the deal.

It has to be noted that there is no specific set of skills that can be programmed. A skilful negotiator should always be prepared, able to steer negotiation constructively, and always be ready to make a reasonable offer. There are a few simple guidelines in addition to the above that should be observed:

- Carefully study all documentation relating to the matter to be negotiated. Consult experts, if there is anything that you do not understand.
- Be punctual.
- Do not volunteer information unnecessarily.
- Keep personal feelings out of the negotiation.
- Remain friendly, even if a deal does not go through.

Analytical skills

Analytical skills refer to an individual's ability to assemble available information and make immediate use of it. The skill for analysis is no different from the formal learning of the problem-solving model. It is the one so popularly used in business management teaching materials, and includes defining the problem, searching for information, analyzing information, selecting alternatives, evaluating alternatives and making a decision. A similar approach can also be used in skills development, and simplified where appropriate.

Most analytical skills involve the use of quantitative methods. Statistical information is all historical, as is accounting information, though more often than not, the latter also includes a prediction of future events such as budgets, forecasts and projections. All historical information must be used with care, since what has happened before may not apply to the future, and any predictions based on historical information has built-in deficiencies, because it includes past inefficiencies into the forecast. Therefore, it should be noted that even the most complete historical information, and sophisticated statistical analysis, still cannot replace human judgement.

There is no short and sweet guide to developing analytical skills, but it is possible to acquire a few simple techniques to enhance an individual's analytical abilities. Those who have no particular interest in quantitative information can learn to accept and appreciate its power by simply using them in accordance with their needs.

For example, percentage expression is more meaningful than figures, and to correlate one figure with another is always fun. In any event, if quantitative analysis is such a drain on an individual's intellectual capacity, it can always be done by computers, or simply by having someone else do the analysis. To be sure, the person doing the analysis should also be prepared to explain the analysis and show how it applies to the particular situation. To play with figures is always fun even if the outcome is wrong, but realize when it is wrong it is a learning experience.

Analyzing human characteristics, abilities and capabilities is not as simple as quantitative analysis. A curriculum vitae will tell part of the story; a reference check and proof of the diploma or degree are part of the input available for analysis. There is also such a thing as gut feeling. Someone once said about analyzing people: "When you see a sincere person, you know it." What else can we say? Yes, there is something more to be said about that. How about the idea of "don't judge a book by its cover"?

Planning skills

To develop an individual's planning skills, the individual first has to learn to like planning. It is not just the skill to plan everything in the head, but to plan in a more formal sense, either by writing the details down, entering them on computer, or recording them on tape. Without formalization, the plan can easily get lost if everything is in your head. Everybody plans, except some plans are more formal than others.

In general, planning skills development include the following:

- Learning to like planning and developing a planning habit.
- Always put the plan in writing, on tape or store it on a computer.
- Learning to appreciate that a formal plan is better than an informal plan, and an informal plan is better than no plan (or just storing everything in the head) at all.
- Remembering that a good plan always has the users of the plan in mind.

As with most skills, development is repetitive in nature, and repetition helps to improve planning skills and develop planning habits. The only caveat is that you must not be a slave to the plan.

Japanese Management and People

It is difficult to superficially understand Japanese management. To build a model from Japanese management and their corporate human resources policy is nothing more than superficial, unless further attempts are made to understand Japanese culture. Only then, can we find out what their management practice is based on. Even so, to model from Japanese management, would at best, make us second-best to the Japanese. In any entrepreneurial inclined corporation, by all means, learn from the greatest, but it is necessary to surpass the greatest, otherwise, is there anything to create and innovate?

While this book makes no attempt to search for the wisdom of Japanese culture, at least one simple fact needs to be re-addressed. In recent years, the Japanese have been going through the same harsh economic ordeals as those experienced in Canada, the US and other Western nations. The large Japanese corporations have dealt with the economic downturn by placing the importance of keeping jobs before profit; unlike "Neutron Jack" of General Electric, who made his name by killing over 170,000 jobs during his tenure as CEO of the company.

At the time of preparation of this book, the Japanese, both at home and the multinationals abroad, were confronted with the lack of entrepreneurial challenge. The fear of not being able to continue the nation's economic growth was written all over the wall. The challenge to Japanese management is no different from the others: while technological advancement led the way to a better means of "producing" and "flow of information", there are few indications of a mindset which cultivates innovative and creative individuals. Japan without a doubt is one of the richest countries in the world, but a Japanese professor said this to the author:

So what if we are rich; our small businesses in Japan and people in Japan are hardly as well off as you perceive. It is the rapid capital accumulation that made large Japanese corporations rich, but they are expanding out, rather than staying in. Our money is out, technologies are out, our well-trained people are out, even medium-sized businesses are expanding overseas as well. Yes, small businesses and people in Japan should be more creative, but how much more creative can we be when social and management challenges are not up to the level of technology.

This Japanese professor's frustration is understandable, and Japan is not alone. The issue was raised long before the technological turning point when electronic technology advancement caught the whole world by surprise. Prince Philip of England warned people that it is the human consequences that we must face due to rapid technological advancement, not the technology itself.

The root of the problem is the departure of microeconomic theory from the need of the common good, and our perceived "market economy", which has failed to close the micro–macro gap (see Figure 14.5). If, on the other hand, we could face reality and recognize that a corporation is not merely a legal entity, but a social entity with responsibility not only to its shareholders but to all contributors of the society, then, and only then, will a corporation become a community of entrepreneurs with the purpose of creating wealth and adding value to the individual, the corporation itself and society.

Figure 14.5   The micro–macro gap and entrepreneurship

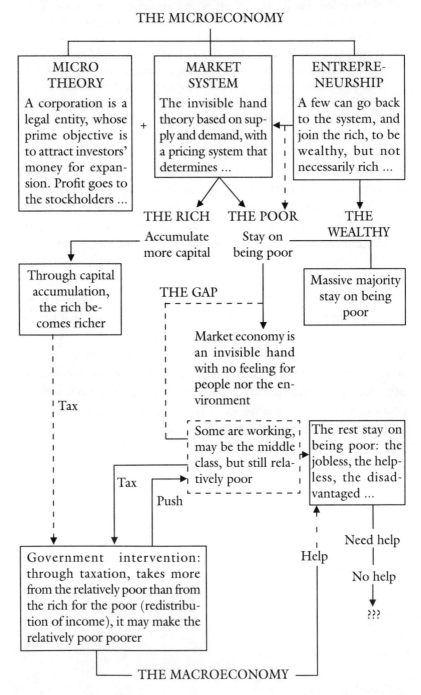

THE MICROECONOMY

MICRO THEORY

A corporation is a legal entity, whose prime objective is to attract investors' money for expansion. Profit goes to the stockholders ...

+

MARKET SYSTEM

The invisible hand theory based on supply and demand, with a pricing system that determines ...

ENTREPRE-NEURSHIP

A few can go back to the system, and join the rich, to be wealthy, but not necessarily rich ...

THE RICH   THE POOR   THE WEALTHY

Accumulate more capital   Stay on being poor

Massive majority stay on being poor

Through capital accumulation, the rich becomes richer

THE GAP

Market economy is an invisible hand with no feeling for people nor the environment

Tax

Some are working, may be the middle class, but still relatively poor

The rest stay on being poor: the jobless, the helpless, the disadvantaged ...

Tax
Push

Need help

Help

No help

Government intervention: through taxation, takes more from the relatively poor than from the rich for the poor (redistribution of income), it may make the relatively poor poorer

???

THE MACROECONOMY

15

The environment and resources: It begins with a bird and its cage

The frontier of the entrepreneurial economy is people and resources. With given attributes and skills, people could use resources to make different combinations to increase their utility function for human use, but if humans exhaust all available resources, we may have to "eat plastic for breakfast, concrete for dinner".

The Bird and the Cage

The following is a quote from Louis Berkhof's *The Doctrine of the Decrees in Theology* (1958, p. 106):

> Man is a free agent with the power of rational self-determination. He can reflect upon, and in an intelligent way of choice, certain ends, and can also determine his action with respect to them. The decree of God, however, carries with it necessity. God has decreed to effectuate all things or, if he has not decreed that, He has at least determined they must come to pass. He decided the course of man's life for him.

Some time ago, a Chinese elder statesman in Beijing explained to foreign visitors the meaning of practising market economy under a socialist system. He said: "Market economy and Socialism are like a bird and a cage; market economy is the bird, and Socialism is the cage. Practising market economy under the socialist system is a simple matter of putting the bird in the cage." At the time, some business friends thought it was a fitting joke about the communist way of doing things: "How can you put a bird in a cage and called it 'market economy'?" But just think twice about what he said; it would not be difficult to appreciate that a similar idea is preached in the Christian doctrine. The doctrine about man's life and his determination and action is much the same as the bird and the cage. Assume that humanity is the bird and the planet is our cage, if the bird consumes and wastes all the food in the cage, without new supplies, the bird would die of starvation. The same could happen to humans. If we exhaust all the supplies that the planet has provided to us, wouldn't we all be dead? Market economy or no market economy, what matters the most is that humanity (the bird) as a whole, and particularly here, the corporate managers, must be realistic and be more sensible about managing resources (supplies) on the planet (the cage), so our supplies can last. Alternatively, we must find new combinations (invention and innovativeness) in the interest of humanity, or move to another planet.

The Concern for Our Environment

Mostafa K. Tolba, Executive Director of the United Nations Environment Programme, said to the Royal Institute of International Affairs, London, on 15 March 1990, in part:

> Were we to continue to waste the natural resources of our planet, new tensions are being created which, unless we, a family of nations, mend our ways, will imperil the political stability of the whole world.
> The great danger is that we will not see the environmental dimensions behind the new conflicts. And we will seek to resolve dispute through arms – a reaction with a millennial pedigree.[1]

These two short paragraphs undermine us on both the issue of the environment and resources waste and utilization. We will deal with the resources challenge from the corporate point of view in brief later in the chapter, but first let us revisit the issue of environmental damage. It may come as a surprise to learn that the world body responsible for environmental health could have claimed that:

Poverty itself pollutes the environment, creating environmental stress in a different way. Those who are poor and hungry will often destroy their immediate environment in order to survive: they will cut down forests; their livestock will graze over grasslands; they will use marginal land; and in growing numbers they will crowd into congested cities. The cumulative effect of these changes is far-reaching as to make poverty itself a major global scourge (World Commission on Environment and Development, 1987, p. 8).

It is without doubt that poverty is a shame to humanity, but the claim that "poverty itself pollutes the environment" by the World Commission on Environment and Development is disturbing. For survival reasons, people in developing countries chop down trees and pollute the environment on a limited scale, but this is far from the damage done by industries.

It is mostly the indiscriminate expansion of capitalism that has caused damage to the environment. It is industries dumping waste, and toxic chemicals in the rivers, lakes, or creating acid rain sometimes making certain parts of the earth unliveable for humans and other living beings. Even tourism – an industry claiming to be a clean and environmental friendly business – its rapid expansion and development have caused damages to the natural environment in Central America, the Caribbean, Southeast Asia and many other countries. Improperly managed waste has also caused environmental concern.

The topic of waste management brings to mind a proposal made in 1996 to dispose US nuclear waste in Canada's north. This "indecent proposal" made the author remember Queen Elizabeth II's speech during her visit to Canada in the mid-1950s; she told Canadians and the world that Canada's north is not only for Canada, but for humanity and the world. It is hard to believe that anybody with more than a basic education could think about using the clean and pure Canada's north as an alternative dump.

As a matter of record, in the United States, during the Reagan administration, there was (and still is) a great concern for the possible long-term impact of environmental regulations on America's industrial future. In particular, some fear that the labyrinth of regulations drives industrial facilities out of business and firms out of the country (Peilly, 1984, p. ix). The following were some of the concerns expressed during that period:

There would be considerable emphasis toward the redirection and growth of chemical process industries abroad. Multinational firms would seriously reconsider their positions with a strong emphasis on foreign investment. Foreign producers, taking advantage of the US situation, undoubtedly would increase capital, R&D [research and development] and related investment in their home countries thus effecting a further transfer of chemical industry growth as well as dominance from US.[2] – A prediction made by the Chemical Manufacturers Association

We would expect environmental pressure to promote a gradual shift of pollution-intensive forms of economic activity from higher income to lower income countries internationally, with the range of activities affected gradually widening over time. This will have notable implications for the development process .... There may be indeed the instance where the export pollution through capital investment abroad becomes national policy in certain economic sectors, to the benefit of both capital export and import countries. – Ingo Walter, one of the first economists to study international implications of environmental regulations (1973, p. 313)

Some 25 years later, the predictions made in the 1970s have become a reality. Multinationals in the US (and for that matter, Japan) have expanded their operations overseas, particularly in the car and chemical industries, either for reasons of avoiding environmental regulatory pressure or taking advantage of cheap labour or both. Moreover, there is the need for the developing countries to attract foreign investment to meet their own economic growth challenges. There is also clear evidence that the massive destruction of forests and the green environment in South America, and Southeast Asia are caused by multinational corporate activities. While the elimination of poverty is a clear mandate for mankind, corporations must assume their responsibility for conservation and the protection of the environment through sensible decisions made in respect to corporate growth and expansion. Unfortunately, a large number of environmental problems are purely managerial decisions, for example, the issue whether to recycle or reuse soft drink containers.

Some 20 years ago, the soft drink manufacturers had a policy to give 5 to 25 cents refund for every glass bottle returned for cleaning and reuse. Now, for no particular reason known to the public, Coca-Cola, Pepsi and all other companies have all switched from reusable bottles to recyclable containers. While this must have a monetary cost benefit to the company, recycling requires a manufacturing process. What is the effect on the environment?

From the point of view of this book, most of the environmental issues related to capitalistic expansion have been discussed in Chapter 5; what we will be doing now is to explore some of the challenges, the responsibilities or measures that fall into the hands of corporate management, and actions that corporations could take for the betterment of our environment.

Sustainable corporate endeavour

The term "sustainable development" or "growth" is used frequently by environmentalists and policymakers when addressing the challenges faced by economic growth on a broader, and larger scale or on a national basis. The concept of sustainable development is also applicable to corporations, particularly those large corporations whose powers to command resources are greater than that of the governments of some smaller nations. In effect, it is a corporation's actions that could help the environment more than any other economic undertakings, because corporations are the dominant force that govern our lives today, and future lives as well.

The mindset

People are generally guided by their mindset which could be changed by external influences or internally generated needs. In the case of dealing with corporate concerns for the environment, there has been some public intervention, such as the OECD's (Organization for Economic Cooperation and Development) resolution to make polluters pay in principle, government environmental protection legislation, and other external forces but these have not made much impact on the corporate managers' decisions.

Attempts made here are essentially a small effort, but hopefully will help to increase awareness towards solutions that could help with the situation. The following are some self-awareness thoughts:

1. Entrepreneurship is a wealth-creation and value-adding process. Merely to create wealth for the individual and add no value to society would not be entrepreneurial. Business actions that damage the environment or cause the misallocation of resources will not add value to society.

2. Care for and conservation of the environment are not the responsibility of the government and environmentalists only, but the responsibility of every individual. In particular, those who have the power to make decisions to allocate resources that affect the environment.

3. Some selfish motives are good motives; for example, our parents raised and provided for us and planned for our future, just as we love our children and children's children, and for those selfish reasons we plan and provide to satisfy their needs. It is perfectly justifiable for us to include ourselves and our descendants in our plans for the future, but our plans for the future must be for the future's sake, not to take what is for the future and use it for our present consumption and enjoyment.

4. Innovation and creativity mean the pursuit of opportunities available to us. Typically, the rate of management innovation is less than the rate of technological innovation and change. Technology comes first, then technology management, followed by management in general. Efforts must be devoted to close the technology and management gap, without taking advantage of technology and destroying the essence of management, or unnecessarily exploiting the environment and resources to satisfy short-term financial objectives.

5. "Caring for the environment and sensibly allocating resources" need not imply increased costs or erosion of the corporation's competitive position. On the contrary, there will be opportunities for product innovation, process improvement, re-tooling and re-engineering in corporate endeavours. The result could create even more wealth for the individual and add greater value to the corporation and society.

6. Accounting for the environment and better resources allocation and conservation could be developed. Development of an accounting method to deal with environmental cost does not

require the approval of the Stock Exchange Commission, or any other authorities. It could be as simple as a set of accounting practices for internal use to guide corporate management and public disclosure, as long as the publication of financial information relevant for "public account" is also exercised.

7. Reuse, recycle or reduce? Where a decision needs to be made, it should be made on the basis of simple cost analysis, taking into consideration the future cost in resources and environmental impact.

Fundamental Operational Principles with Concern for the Environment should be Observed by Corporate Managers

Over the years, rules which outline ways to protect the environment have been developed by various groups or individuals to guide people, and the rules are particularly of interest to corporations seeking expansion. These rules may lack enforcement power, but they do help to guide managers in making business decisions. Some of these rules (Turner et al., 1994, pp. 59–61) with added notes from the author are listed:

1. Market and intervention failures related to resources pricing and property rights should be corrected. (Note: Public intervention in land pricing, and pricing of very scarce resources should remain.)

2. Maintenance of the regenerative capacity of renewable natural capital (RNC), i.e. harvesting rates should not exceed regeneration rates and avoidance of excessive pollution which could threaten waste assimilation capacities and life systems. (Note: Managerial action to prevent public intervention is needed. For example, sewage treatment plants are designed to deal with human, not industrial, waste. Toxic chemicals must not be dumped into the public sewage system, or indeed into the ocean, lakes, rivers or streams. Corporate actions to deal with the situation could include 24-hour sampling of waste water, using automatic samplers; regular analysis of samples; and 24-hour monitoring of the volume of waste water.)

3. Technological changes should be steered via an indicative planning system, encouraging switches from non-renewable (NRNC)

natural capital to RNC. Efficiency-increasing technical progress should dominate. (Note: The use of natural rubber versus petroleum-based products reduces the reliance on NRNC – both can increase efficiency and product quality.)

4. RNC should be exploited, but at a rate equal to the creation of RNC substitutes (including recycling). (Note: R&D in material science, exploiting the use of agricultural products to replace chemical products should be encouraged. For example, cotton and wool products are still superior to synthetic materials. In order to support agricultural products, efforts could be developed to support an agri-industrial based economy taking into consideration the preservation of a clean rural environment.)

5. The overall scale of economic activity must be limited so that it remains within the carrying capacity of the remaining natural capital. Given the uncertainties present, a precautionary approach should be adopted with a built-in safety margin. (Note: Corporations should work closely with local environmental agencies and ecosystem specialists to determine the sustainability for corporate growth. Stress on a planned sustainable corporate growth, rather than the simple growth strategy driven by profit and the word too often used as an excuse: "competition".)

On the basis of the above principles, R. K. Turner (1994) prepared his practical solutions which can be used to guide corporate managers' actions (see Figure 15.1).

This illustration is intended to be a reflection of current thinking, and is not universally applicable to corporate managers. Nevertheless, it is a good general reference to guide decisions.

Essential Concepts for Achieving Positive Environmental Attitude

In addition to the above mindset reminders, and some fundamental principles in respect to the preservation of environmental health, there are numerous relevant concepts which have been developed since the early 1940s. Recently, Michael Atchia (1995, pp. 7–9) packaged them into 24 essential concepts designed to achieve positive environmental attitudes. They are as follows:

Figure 15.1 Sustainability practice

Sustainability mode (overlapping categories)	Management strategy (as applied to projects, policy or course of action)	Policy instrument (most favoured) • Pollution control and waste management • Raw material policy • Conservation and amenity management
VWS (Very weak sustainability)	Conventional cost-benefit approach: correction of market and intervention failures via efficiency pricing; hypothetical competition; consumer sovereignty, infinite substitution.	e.g. pollution taxes, elimination of subsidies, imposition of property rights.
WS (Weak sustainability)	Modified cost-benefit approach: extended application of monetary valuation methods; actual compensation, shadow projects, etc.; systems approach, "weak" version of safe minimum standard.	e.g. pollution taxes, permits, deposit-refunds; ambient targets.
SS (Strong sustainability)	Fixed standards approach: precautionary principle, primary and secondary value of natural capital; constant natural capital rule; dual self-conception, social preference value; "strong" version of safe minimum standard.	e.g. ambient standards; conservation zoning; process technology-based effluent standards; permits; severance taxes; assurance bonds.
VSS (Very strong sustainability)	Abandonment of cost-benefit analysis; or severely constrained cost-effectiveness analysis; bioethics.	Standards and regulation; birth licences.

Source: Turner et al. (1994, pp. 59–61).

1. The size and range of population is regulated by available physical resources (e.g. space, water, air, food) and by biological factors (e.g. competition), hence, the concept of carrying capacity of a natural environment.

2. The flow of energy and material through an ecosystem links all communities and organisms through a complex of chains and webs that invariably starts with plants.

3. The planet is made up of a number of interacting and interdependent components.

4. We live in a "spaceship-earth", a closed system characterized as having limits.

5. Human activities and technologies influence considerably the natural environment and may affect its capacity to sustain life, including human life.

6. A mode of life heavily dependent upon rapidly diminishing non-renewable energy sources (e.g. fossil fuel) is unstable.

7. The relations between humans and their environment are mediated by their culture (their mode of life and habits).

8. A clear difference exists between the natural needs of human beings and those wants and desires artificially created by advertising and social pressures.

9. Economic efficiency often fails to result in conservation of resources.

10. Rational utilization of a renewable source (e.g. rate of fish caught equal to rate of natural regeneration of fish population) is a sensible way of preserving the resource while obtaining maximum benefits from it.

11. Sound environmental management is beneficial to both man and environment.

12. Use of resources demands long-term planning if we are to achieve truly sustainable development.

13. Elimination of wastes through recycling and the development of clean technologies are important to modern societies to help reduce the consumption of resources.

14. Certain article contaminants (e.g. radioisotopes, mercury, DDT) are too long-lived or of such a nature that natural processes are unable to eliminate them readily.

15. More often than not, the ocean has become the dumping place for chemicals, oil, sewage, agricultural waste, etc.

16. Wildlife populations are important aesthetically, biologically (as gene pools), economically and in themselves.

17. Nature reserves and other protected wilderness areas are of value in protecting endangered species because they preserve their habitats.

18. Destruction of natural habitats by human beings is the single most important cause of extinction today.

19. Destruction of any wildlife may lead to a collapse of some food chains.

20. The survival of humanity is closely linked to the survival of wildlife, both being dependent on the same life-support system.

21. The protection of soils and the maintenance of sustainable agriculture are essential factors in the survival of civilization and settlements.

22. A vegetation cover (grass, forest) is important for the balance of nature and for the conservation of soil, besides being an exploitable natural resource.

23. Soil erosion is the irreversible loss of an essential resource (and must be prevented).

24. Cultural, historical and architectural heritage are as much in need of protection as is wildlife.

Environment Accounting

"Accounting is the language of business" is a simple expression given by accounting teachers to their students encountering for the first time the challenges of learning accountancy. The idea is simple enough to understand; accounting reduces everything into a single denominator, a quantitative expression in the currency of the nation. Unfortunately, while this creates uniformity in accounting practice, it only deals with transactions with material content, which is the fundamental requirement from stock exchanges for listing companies. For example, even though psychological loss cannot be measured accurately in dollars and cents, it must be measured if a lawsuit is involved. One case in point is the breast implant damage claim made against Dow Corning (see Chapter 8); a victim was awarded US$25 million in damages. The figure of US$25 million certainly could not have arisen from any accurate measurement of pain and suffering. Nevertheless, the jury made a decision, and it is now embedded as a legal precedent. Once, during the 1941 Japanese bombing of Chungking, the author found himself in an air raid shelter for the better part of a day (this was not unusual – the Japanese bombing was continual, ruthless and opposed by only a token air defence) and ended up having to relieve himself in the street. He was caught by the civil defence. The penalty? Removing corpses from the burning streets for the rest of the day.

These two stories illustrate one basic concept; so long as someone is in a position of power anything can be assigned a value. Just as the jury determined the damage to the woman to be worth US$25 million, in the case of the author's personal experience, the cost of polluting the street was to carry corpses to a piling ground at a central point.

There is no recognized way to account for the author's experience. As a teenage refugee, I had no opportunity cost, nowhere to complain, and whether anybody cared was totally irrelevant. However, in the case of the Corning suit, the accounting entries can be made for:[3]

- The first entry required is to establish the lawsuit, immediately upon becoming aware of the situation:
  Cr. Provision for lawsuit
  (or other fitting account) $25,000,000*
      Dr. Accumulated earned surplus $25,000,000
- The second entry would record the actual award ordered by the court to pay the woman (or her lawyer's trust account):
  Dr. Provision for lawsuit
  (or other fitting account) $25,000,000
      Cr. Cash $25,000,000
- * This is an assumed amount, used for illustration purposes only. It could be much greater than $25,000,000 or less.

Under the circumstances, it is, perhaps, better to leave the accounting for the woman to a professional, if necessary.

In the Corning case accounting entries could be easily made because the cost (sacrifice) to the company was measured precisely for $25 million. But what would it be for environmental damage?

And who is going to determine what? There is no clear evidence that accountants have openly undertaken the task, but it is not impossible to do so. The following are some examples to support the claim.

In 1975, the Organization for Economic Cooperation and Development (OECD) published *The Polluter Pays Principle*; in three principles, it states:

> The principle to be used for allocating costs of pollution prevention and control measure to encourage rational use of scarce environment resources and to avoid distortions in international trade and investment is called "polluter pays principle". This principle means that the environment is in an acceptable state. In other words, the cost of these measures should be reflected in the costs of goods and services which cause pollution in production and/or consumption. Such measure should not be accompanied by subsidies that would create significant distortions in international trade and investment.

Accordingly, the cost in environmental prevention borne by the corporation and any damage to the environment should be calculated and included in the cost structure as part of goods manufactured. However, this involves two sets of calculations; one is out of pocket cost which can be accounted for with no difficulty, while the second considers future costs and is the real challenge. The cost may not involve cash disbursement by the company. For illustration purposes, we will use Professor T. M. Das' estimate of the value of destroying a 50-year-old tree (see Chapter 7). Let us use a disposable chopsticks manufacturing company's cost sheet. Bear in mind, the largest user of disposable chopsticks are Japan and China, but now other countries have also developed the habit of using them.

- Assumption 1. That the accounting income calculated under the current state is based on "Generally Accepted Accounting Principle", where the chopstick manufacturer's financial accounting is no different from other practices'.

The following are sets of financial data based on the normal production and normal costing conditions.

- Assumption 2. For the purpose of illustration, two sets of wood-costing methods are involved; the conventional costing based on normal recording including cutting, transporting and processing trees to make the finished products, and the second set of costing, based on Professor Das' estimated value of a 50-year-old tree, proportionally according to the age of the tree.

- Material: soft wood. The time it takes a tree to mature is about ten years. A tree of good quality could produce 10,000 pairs of chopsticks. In accordance with operational records, it averages per two trees = one good tree = 10,000 pairs of chopsticks. Current cost of cutting down a tree is about $10 per tree, which means $20 = 10,000 chopsticks.

- Financial income: based on a standard volume of producing and selling 1,000,000,000 pairs of chopsticks, the material cost for wood alone would be $2 million. All other costs are $1,000,000. Further assume the sales for a particular year was $4,000,000, the financial income accordingly was $1,000,000.

Based on Professor Das' estimation of the value of a 50-year-old tree as $194,250, then a ten-year tree is estimated at approximately $20,000 per tree. There were 200,000 trees used to make 1,000,000,000 pairs of chopsticks; therefore, the revised material cost for making the above amount of chopsticks would be:

$20,000 × 200,000 trees = $4,000,000,000

Revised income (based on the environmental value of a tree) would be:

Financial income	$ 1,000,000
Add back conventional cost estimate for 200,000 trees × $20	4,000,000
Financial income before deducting tree cost	5,000,000
Deduct tree cost based on environmental value	400,000,000
Revised loss	$395,000,000

Shocking! Do we still want to use disposable chopsticks? For that matter, the same goes with the massive destruction of forests for cattle grazing. What would be the environmental damage cost of a hamburger based on similar costing? Do we still believe in mass selling of hamburgers? If so, how much will the environmental cost be? Who bears this cost, since there is no way anyone can sell one hamburger for $16,000+ (a rough guess based on the cost of disposable chopsticks)?

If an accounting entry is to be made for the cost of chopsticks, it would look like this:

Dr. Cost of goods sold	$395,000,000	
Cr. Sales	$ 4,000,000	
Cr. Environmental damage	$391,000,000	

Where does the environmental damage go? It cannot be an item under the equity section, but a liability, which simply means we have borrowed from the future's life-support system that we can never repay.

Is there any solution to the situation? Purely from an accountant's point of view, the answer is yes. If the Dow Corning case gives the idea of what a wrong management decision can do to the company's future, the same approach could also be used to help to remedy our indebtedness to our environment. The method involved need not be as complicated as the Dow Corning case; we could make provisions for the environmental damage and anticipated environmental cost yet to materialize. This is illustrated as shown in Figure 15.2, with the assumption that environment damage has a direct linear relationship, such that the environmental cost is calculated to be 10% of current earnings.

In order to confront the potential problem of environmental damage, a 10% special provision based on current earnings is to be provided, and the accounting record could be shown as:

To provide for environmental care 10% of current earnings:

Current earnings $5,000,000 × 10% = $500,000

Special provision for environmental care	Accumulated retained earnings
$500,000	$500,000

When an expense is incurred for environmental care, charges can be made out against a special provision for environmental care.

Dr. Special provision for environmental care $100,000

    Cr. Cash $100,000

Figure 15.2   Environmental cost of a corporation's expansion

(a) Assumption: A linear relationship exists between the rate of corporate growth and the rate of environmental damage

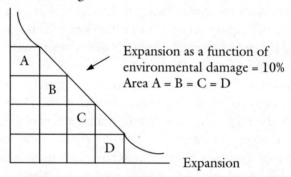

or

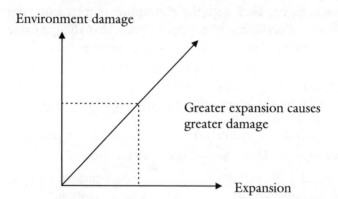

Figure 15.2 (cont'd)

(b) Excessive expansion imposes even greater burden on the resources and accelerates rate of environmental damage

Burden on the resources, accelerate
environmental damage

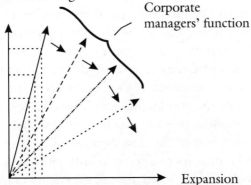

(c) Sustainable expansion (caring for the environment and sensible allocation of resources) unavoidable impact on the environment

Unavoidable impact on the environment

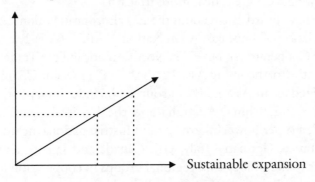

It should be noted, there is no tax implication since the provision is made against accumulated retained earnings, not from the financial income.

Please note that there are many other ways to do the accounting, including charges made directly against revenue, for calculating operational residual and value-added purposes, which will be dealt with in Chapter 16.

Island Economy Versus Continental Economy and Resources

Our earth is made up of water and land, and humans are inhabitants of the land consisting of islands and continents. Island economies are based on a narrow set of activities, which are highly susceptible to external economic influences. Continental economies, on the other hand, can accommodate a greater range of activities. World economic growth must rely on economic cooperation rather than confrontation between the two. Island economies depend on the continents' resources for economic growth and expansion; while the continental economies do not need the islands, island economies provide unique environments, both in the ecological sense, and in the development of a variety of human activities and industries can occur in both directions. Unfortunately, if history is to be believed, the sad experience of humanity is that confrontation is as common as cooperation, if not more. The end of the Cold War resulted in much optimism on this score, but events since then (in Bosnia, Ethiopia, and many other places) show we have a great deal more to learn.

We have heard much about the EU (European Union), ASEAN (Association of Southeast Asian Nations), APEC (Asia-Pacific Economic Cooperation), NAFTA (North American Free Trade Agreement), the forthcoming Asia-Pacific "G-6" (a recent US initiative, organized as an Asia market group to include Singapore, Japan, Hong Kong, China, Australia and the US itself, pairing with "G-7", a super powerful group of industrial nations including the US, France, Germany, Italy, UK, Canada and Japan), and other smaller trading blocs, and regional groups in cooperation for economic development. What are all these economic cooperations about? Perhaps the European (EU) model can tell us a lot as it has an excellent spirit of cooperation, beginning with agreements on resources, trade, nuclear bans, defence and eventually political union. It still has plenty of problems. To this day, even though the Euro-currency has been tested in France, and the rate of exchange for member nations must be determined and finalized by the year 1999, the UK government refuses to submit the pound sterling to the control of any other exchange media. Despite these difficulties, the EU has loftier goals than any other agreement, union, or cooperation in its attempts to eliminate regionalism and re-strictions on all economic activities, including movement of la-bour. "Free trade", a beautiful term that signifies the free movement of all resources, at least among EU members, is becoming a reality. The results are difficult to judge, but it is clear that if "freedom" means anything to humanity, "labour" must be allowed to move freely as well. This is extremely difficult for many nations. How do we address the question of cheap labour invading our market and eroding our living standards? How are we going to monitor our population growth? What sort of socio-economic problems will it create for us? How can people go freely around the world without a passport, visa or special permit? Remember, it will be a special privilege for Hongkongers to travel to the UK without a visa after 1 July 1997, but there is still the problem of a passport, and forget about getting a job in the UK. And this after more than 100 years of being His or Her Majesty's loyal subjects? Why? Can we ever imagine nations without customs and immigration controls? What would the world be like, should this be the norm rather than the exception?

Labour movement is not the only challenge, there are also re-sources challenges. This is especially true in the island nations. As most islands have no meaningful resources; they cannot create monopolies to operate large-scale operations, or develop substan-tial internal markets; nor can they raise a large amount of capital/finance on the home markets. For capitalistic expansion, corpora-tions must expand beyond the island nations to where they can find the three essential factors: land, labour and resources. For this they must move to the continents. But economic growth even on the largest continent puts pressure on resources; therefore, for the corporate manager, the challenge is not the traditional idea of "marketing, marketing, selling and selling". The challenge is how meaningfully and sensibly the managers can manage the resources. We do not expect to have "resources wars" in the future, at least if the "World Trade Organization" is going to work. (Note: Resources-rich countries include Australia, Canada, the US, China and some continental countries and sub-continental countries such as India and Russia; the control of resources could be a national economic strategy.) But for the individual corporation, managers must bear the responsibility of seeing that the resources are not wasted, and efficiently utilized. For this reason, the bottomline approach to corporate management must step down and allow the use of an engineering costing approach (ROI) for resources allocation, which is based on a set of principles. The following are a few suggestions:

1. Determine the opportunity cost of resources employed by the corporation. In other words, the efficient use of resources for other products. For example, look at why others use the same resources to make different products.

2. Based on opportunity cost, study and determine the theoretical cost of the product, without giving a standard allowance.

3. Quality control must be extended to the control of environmental factors. That is, no quality control is complete without taking into consideration environmental factors, particularly when dealing with industrial waste.

4. Use renewable resources if possible.

5. Search for materials that decompose quickly and without causing environmental damage. For example, certain plastic materials will take approximately twenty years to completely decompose and are extremely unfriendly to the environment. There are other materials with better environmental health qualities such as paper and cardboard.

6. Promote the reuse concept and devote design of the product's container/holder for the same purpose. For example, glass food containers come in a variety of sizes and quality, many of them are difficult to clean and designed for disposal. For the purpose of a healthy environment, the design can be changed to a standard wide-mouth for easy cleaning, thus promoting reusability; a deposit box for returnables can be designed to reduce breakage.

7. Develop two sets of capital cost (depreciation) charges, one for natural capital, and the other for capital created through human efforts.

8. Natural capital cost must be determined, including water, clean air and soil (not the cost of land, but the quality of soil).

9. Use a theoretical material cost system; cost must include cost to the environment.

10. Use cost (determined in no. 9 above) based on pricing, adjusted to demand and competition.

16

Bitter sweets: The corporate performance

There is no excuse for us to say that we, individually or collectively are not aware of our corporate challenges about caring for the people, environment and resources – what with all the commotions and appeals made by environmentalists, and kind hearted individuals' concern about human relations in the industry. Economists have done their part by building mathematical models to account for the environment cost and behavioral scientists have advocated various ways and means for the caring of people. But so far, neither have made any impact for accountants to make the effort to incorporate these "hidden costs" into the accounting models so that the models can work for the corporate management and humanity.

This chapter attempts to demonstrate what can be done by employing accounting methods to pave the way. Hopefully, works of this nature will continue, so that fruitful solutions to the challenges we, particularly corporate managers, face, can be found.

The Judgement Day

There is always a day of judgement. As a person, we have been and always will be judged by others on our performance, as we judge others on their performance. All judgements are based on certain yardsticks, social criteria or, in extreme cases, moral or legal considerations. A corporation, if it is viewed as a legal entity, is judged by its bottomline performance. In a profit motivated culture the attendant notion is that rewards must always be linked to profits.[1]

329

But profit is an ambiguous word, causing distress to economists and accountants alike. Profit provides the motivation that pushes managers to strive for short-term artificial performance, and more often than not to sacrifice long-term benefits, not only for people and the environment, but even for corporate well-being. One striking point was made by Mr Wee Cho Yaw, the CEO of UOB Group. In his address to an audience of academics and business persons, the successful bank executive said: "In the West, millions are being paid to professional CEOs to manage multinational corporations. These are well-educated and very experienced people." "But again," he said: "since their benefits and their bonuses are mainly based on current profits, they would be more inclined to go for short-term gains by taking bigger positions even if this means taking on greater business risk."[2]

"Profit", for better or worse, has weathered the storms of historical controversy; it is the crown prince to some, but profanity to others. It lacks conceptual clarity, and how we measure it is, if not hogwash, at the very least, a mess, but in today's business world, it is all we have. Could it be that this is the fundamental truth in life? Or have simply we been spoonfed, and spoiled, with no wish to break the knowledge barriers? Nevertheless, there had been numerous attempts made on "profit's life"; Karl Marx did it, so now Raymond W. Y. Kao does it, except he does not intend to kill "profit", but wishes to search for its true meaning. Neither does he disagree with having profit, except he just wants to know what profit is, and what it does to people and the environment. Make profit by all means, except it should be real profit, and not at the cost of people and the environment.

Here at the judgement point of this book's deliberation, let us put "profit" out of circulation, and try to explore any alternatives. Yes, there are plenty, provided by the accountants for one. A large number of accountants choose not to use the traditional bottom-line "profit", rather they use "financial or accounting income", so do others, including Anthony, Dearden and Vancil, Stevenson, a whole lot of the Harvard scholars and others advocate the idea of "residual income". Today, in celebration of the writing of the last chapter of this book, allow me to make an attempt to break a small knowledge barrier. Let me introduce the use of the "added-value" or "value-added" approach to measure a corporation and its managers' performance, not on a short-term, but on a long-term basis. Then, there is the question of "how do we survive, if our shareholders' ROI interest is not clearly attended?" The answer is quite simple, the value-added approach does not kill the traditional notion of "profit". By all means, calculate "ROI" based on the way it used to be, except the value-added approach will give a more meaningful measurement. The decision to invest in certain corporations will always be the investor's decision, but, will be guided by how stock analysts express their analytical opinions of the trustworthiness and well-being of the corporation, measuring the corporation's performance as a community of entrepreneurs created for the purpose of creating wealth and value for the individual and the common good.

The Three Kings from the Orient

There is the Messiah on business; the three kings from the Orient bear the gifts of "financial income", "operational residual" and "added value".

Financial income, the first of the three kings

We need not explain what financial income is, but there is a difference in the entrepreneurial approach to deriving financial income. The corporation could follow the traditional way of determining corporate financial income, and simply make a name change; instead of "profit" call it "financial income" or "accounting income". It is expressed as:

Accounting or financial income = Total revenue – Total cost

For illustration purposes, let us assume that the Earth Corporation has earned an accounting income of $1,000,000 which is about 10% of sales; its sales would thus be $10,000,000. Using the above illustration, we could express the operation by showing the following:

Revenue $10,000,000 – Cost $9,000,000
= Financial income $1,000,000   (1)

The measurement of revenue and cost is the expensed capital (or assets) derived on the basis of generally accepted accounting principles (Kao, 1986), satisfying auditing guidelines and practising requirements. Alternatively, the following could be considered.

Operational residual, the second of the three kings

Operational residual is the business income determined by accountants and is closer to the economist's idea of "profit". It is a simple matter of recognizing the investors' entitlement. In other words, rather than considering investors' entitlement to include all profit with all its uncertainty, because the corporation may not earn a period profit at times, capital investment is recognized as requiring its rightful return. Just like labour is entitled to wage payment, money lenders receive the agreed interest, and landlords receive their rent, investors ought to have their payment (much the same as interest) for capital service, namely a capital charge. This is an accepted practice. Using the earlier example, assume that the Earth Corporation has a capital investment (in the form of equity) of $2,000,000, and market interest based on term loan lending rate is 8% per annum. The investors' capital charge to the Earth Corporation should be: $2,000,000 × 8% = $160,000. Hence, operational residual of Earth Corporation could be calculated as follows:

Financial income (calculated in equation (1) earlier) $1,000,000
  – Investors' capital charge $160,000
  = Operational residual $840,000

If the Earth Corporation has already implemented the stakeholder's scheme, and assuming that the stakeholders' recorded stake is $200,000, and capital charge equal to the charge made in favour of stakeholders' stake entitlement or 8%, then a further $16,000 capital charge should be made, thus reducing operational residual from $840,000 to $824,000.

Added value (or value added), the last of the three kings

Added value is a beautiful concept that has been effectively used by Southeast Asian countries (the little dragons) to make them economically strong nations. It is a simple import–export manipulation: import goods requiring additional manufacturing process at a value below that of finished goods, add value (through labour, technology, and/or additional materials) and re-export with higher value to other countries, including the country of origin. Often, that country is the US. These value-adding countries are characterized by:

- No meaningful resources of their own
- Highly national motivation for economic achievement
- Relatively stable government
- Strong currency
- Trade surplus

The value-added concept practised by these little dragons demonstrates the economic reality that reflects the Decrees in Theology:

> Man is a free agent with the power of rational self-determination.... God has decreed to effectuate all things or, if he has not decreed that, He has at least determined they must come to pass. He decided the course of man's life for him (Berkhof, 1958).

It is clear, since "He has at least determined they must come to pass" and we are unable to add any resources to our planet, that the economic reality is that anything we do will be to make different combinations to increase the utility function for our use. If so, where is the "profit"? Ultimately, it can only be viewed with respect to gains to our mother earth. Consequently, the value-added approach to measure corporate performance can be made meaningful to humanity as a whole, as well as to the individual and the corporation.

Under the value-added approach, a corporation needs to measure the scarce resources beyond the exchange value, and the damage done to the natural environment, either accountable as part of the cost of goods sold, or as a deduction from the traditional "profit" or retained earnings, because they are not earnings, but part of the life-support system that belongs to the future.

For illustration purposes, the tourism industry is used as an example. Assume a hotel chain, a spin-off of the Earth Corporation, has acquired a permit to build a 300-room resort hotel somewhere on a South Pacific island. It would incur the following costs:

Soft cost (preparation costs, including entertainment, obtaining permit, licensing, professional fees, etc.)	$2,000,000
Estimated construction cost (finished and fully furnished, 300 guest rooms @ $250,000 per room, including all service facilities)	750,000,000
Total estimated cost (one day before opening for business)	$752,000,000

For convenience, the above cost will be amortized over a 20-year period. "Human capital cost" determined as above will be charged against revenue in the amount of: $752,000,000/20 years = $37,600,000. Assume that building the hotel will result in changes in the environmental balance, directly including the loss of trees, and the damage to the coral reef, and indirectly, loss of marine life, and other living creatures. Some damages are permanent, others are temporary, but there are other positive factors that will be considered as improvements in general. However, the change to the environmental balance or damage made to the environment is estimated over a 20-year period to be a grand sum of $200,000,000, or $10,000,000 per annum (which means it may take 20 years for nature to restore its balance either here on the island or elsewhere). To recognize the change or damage, an amount of $20,000,000 will be charged as natural capital cost against earnings. Consequently, the following are required:

One time entry
Date: 1 July 2019

Dr. Natural capital	$200,000,000
Cr. Long-term environmental liability	$200,000,000

On an annual basis
Date: 31 July 2020

Dr. Natural capital expired cost	$10,000,000
Cr. Natural capital	$10,000,000
Dr. Operating profit (or retained earnings)	$10,000,000
Cr. Long-term environmental liability	$10,000,000

Based on these provisions, the Earth Hotel and Resort Inc.'s year-end performance is:

As of 31 July 2020

Assumption 1. Shareholders' equity	$1,000,000,000
Market long-term lending rate	8%
Assumption 2. Stakeholders' equity	10,000,000

Financial Summary as of 31 July 2020
Financial Income as per Income Statement
    for the operating period 31 July 2020

After tax	$100,000,000
Capital charges:	
Shareholders' entitlement:	
$1,000,000,000 × 8% =	$80,000,000
Stakeholders: $10,000,000 × 8% =	800,000
	80,800,000
Operating Residual	$19,200,000
Less: Natural capital expired cost	10,000,000
Value Added	$9,200,000

To summarize the above, here are the three kings:

Financial income (after tax)	$100,000,000
Operating residual (reflecting shareholders' and stakeholders' entitlement)	$19,200,000
Value added to the entity and society	$9,200,000

Financial Ratios

Rate of return on shareholders' investment – based on traditional legal entity concept $1,000,000/$10,000,000	10%
Operating residual as a percent of shareholders' investment $19,200/$10,000,000	Approx. 0.2%

Value added to society (after provisions
made to restore environmental balance
and damage $9,200/$10,000,000)   0.1%

In the preceding example, the corporation and its managers measured the "three kings" on a long-term basis. Why? Because the
measurement not only provides three aspects of corporate performance, but also has elbow room, namely cash conservation of
$10,000,000 charged against operating profit or retained earn-
ings without cash disbursement. However, if the decision of shareholders is to use $10,000,000 as dividend in addition to capital
charge, then, return per share would be:

Capital charge (shareholders' entitlement)	$80,000,000
Natural capital cost expired	$10,000,000
Total	$90,000,000
Shareholders' claim	$90,000,000/$100,000,000 = 9%

The $90,000,000 shareholders' claim is no different from the tra-
ditional "profit" allocation, other than the amount of $800,000
allocated for the stakeholders' share of the harvest. The difference
is that the shareholders are made aware that an additional
$10,000,000 representing the "put something back" to the envi-
ronment for what damage the corporation's operation has caused,
is now redistributed to the shareholders as their additional "entitlement" for their investment.

What should the Managers Do with the Cash?

As $10,000,000 charged against operating profit or retained earn-
ings is, in fact, an entry for the accounts without cash disbursement, the managers could, therefore, use their discretion to launch
a series of programmes to restore the environment or similar purposes. Just as an example:

Restoring the environment:
Donation to environmental research	$1,000,000
Restoring coral reef	1,500,000
Reforestation	1,000,000

Prevent land degeneration	1,000,000
Coral reef maintenance	500,000
Restoring fish stocks	500,000
Wildlife protection	500,000
Tourists and employees' education in environment care	1,000,000
To create Earth Care Foundation associated with corporate growth and expansion	3,000,000

Bearing in mind, all expenditures are tax-deductible, therefore, an income tax effect of $5,000,000 can be used for other meaningful conservation undertakings.

All of the preceding examples are merely for illustration purposes, the figures are not real, nor is the situation. The purpose of the example is to explain how it is possible, and by so doing, show that it does not in any way affect the corporation's performance, because in so far as "profit" is concerned, the reported financial income should be exactly the same as before. Stock analysts, if they so wish, based on the reported financial income could analyze the corporate performance and derive their conclusions the same as before. With the exception that, the additional people and environmental care information should provide the added value to the corporation, therefore, making it more favourable for serious investors investing their money on a long-term return basis.

Corporate Reporting

Under the legal entity concept, a corporation is responsible to its shareholders. It goes without saying, that any human being working in the corporation is an employee. As a corporation's employee, he or she is responsible to the shareholders as well. On the other hand, under the social entity concept, a corporation is a social entity responsible for its actions to all contributors, including shareholders, plus all others, and those working in the corporations are entrepreneurs rather than employees. The corporation is, therefore, a community of entrepreneurs. Individuals in the corporation are responsible to themselves, working to achieve two common purposes:

- To serve people
- To trust others and earn others' trust

This is business, entrepreneurial business.

Traditionally, because a corporation is responsible to its stockholders, the entire exercise of reporting, other than to satisfy legal requirements, has been only to concentrate on pleasing the stockholders and tell them how much money the corporation made in the last year, what kind of dividend they can expect to receive, what the major investments are, and use every means possible, to satisfy the stockholders. Under the social entity concept, a corporate report will also discharge its responsibility as a corporate citizen for its value-added undertakings. Based on these, a comparison between considering the corporation as a legal entity and a social entity is made in Table 16.1.

In addition, under the social entity, it is also feasible for the CEO to summarize corporate initiatives which care for the people working in the corporation. The initiatives are:

- Lifetime employment plan
- Stakeholders' entitlement of harvest
- Employees' equity share participation plan
- Employees' family protection plan
- Employees' lifetime care plan
- Pension plan with the corporation's contribution and schemes of transfers as well as flexible withdrawal for important purposes

The comparison in Table 16.1 is merely intended to project the differences between viewing a corporation as a legal entity as opposed to a social entity. As the corporation does not abandon earning "financial income (profit)" as being its financial performance, the operational residual and added-value approach could only enhance the corporation's commitment and public image. It should be clear that none of the above is intended to disregard stockholders' interest. If nothing more, the practice of imposing a capital charge on shareholders' investment is to assure the basic earnings for shareholders' investment, unlike the traditional "profit"

Table 16.1 Corporate reporting: Legal entity versus social entity

Reporting	Legal entity	Social entity
Financial review	Revenue, profit, share value, share capital, long-term debt to equity ratio	Same as legal entity
Dividends	Declared, paid	Capital charge = dividend, dividend paid to both shareholders and stakeholders
Operational review	Assets management, business strategy, competition analysis, market penetration, among other things, directions in forthcoming years	Same as legal entity + stressing corporate purpose and its philosophy about people, environment, shareholders' interest and those of other contributors
Corporate review	Corporate status, change in corporate organizational strategy and change in capital structure	Same as legal entity + corporate image, corporate concerns for the environment and necessary steps taken (or planned to take) to protect environment and corporate commitment to sensibly allocating resources and conservation
New initiatives	Reporting for initiative taken by the management	Initiatives taken by individuals in the corporation, and management commitment to develop corporate enterprising culture. Initiatives about employees attributes and entrepreneurial skills development as well as adopting the use of multiple measurement for corporate performance, including the financial income, operational residual and value-added approach for corporate reporting

Table 16.1 (cont'd)

Reporting	Legal entity	Social entity
Directions	Protect stockholders' investment and attempts to improve earnings	Continuing the pursuit of opportunities, management committing to develop corporation as a community of entrepreneurs. Continuing efforts will be devoted to earn people's trust in the marketplace, as well as from the corporation's employees
Resources and future	Greater emphasis placed on acquiring cheaper resources and up-to-date technology; attempts will be made to reduce labour cost and overall cost improvement to realize better earning potentials	Sensibly allocating resources, prevent waste and eliminate if possible damage to the environment. Develop new resources supply proven to be environmentally friendly, and the use of engineering approach to cost determination. Theoretical costing method will be used with no standard allowance
Expansion plans	Under the competitive environment, business expansion is a necessity not just to be the best in the business, but also for defensive purposes	People and environment before expansion
Public relations	Emphasizing we offer consumers' the best product lines, and we are the best in the business	Same as legal entity + the corporation's commitment to people, environment and will do what it can to earn the trust of people

approach whereas shareholders' have no basic earnings entitlement. This, in fact, would reduce some elements of uncertainty as a result.

CEO or Directors' Report

The CEO and/or directors of the corporation are in the highest management positions, leaders are given the position by using the corporate by-laws, normally elected by individuals who have the majority shareholdings. In the highest position of the corporation, the CEO should highlight new corporate initiatives, particularly those with long-term benefits to the corporation and society. He or she should clearly project corporate commitment, with performance detailed in the corporate report as shown in Table 16.1 under the social entity concept for corporate reporting. The suggested report from the CEO of an entrepreneurial corporation could include the following:

EARTH CARE FOUNDATION

The Earth Hotel and Resort Inc.'s performance in the past year has set the stage for its continued growth in the people focused industry, while at the same time devoting efforts to preserving and improving environmental health. The foundation will propel the company's growth, to develop new people business with a commitment to serving people, and through our dedication and unquestioned devotion to our clientele to earn the people's trust, but not at a cost to our environment, if at all possible.

Our new core business in hotel and resort development includes a direct linkage with the need for a healthy earth. It is a blend of concern for people and the earth that captures our enterprising spirit to build a foundation for users in the years ahead, not only for ourselves, but for our children and our children's children as well.

REPORT FROM THE CEO

Earth Hotel and Resort Inc.'s performance in 2020 has been good. There was a significant increase of 211% in financial income after tax attributable to shareholders, from xx million to xx+ million. This represents an increase in earnings from 10.2% to 21.1% per share.

Our shareholders and other members of the corporation will be pleased to know that as a leader in our business, we have taken the initiative to implement our stakeholder scheme providing an opportunity for all our employees to establish a stake in the corporation. Moreover, as our business is essentially a people business, and directly bearing on our relationship with nature, the Board of Directors empowered me to initiate a new accounting approach to measure the impact made to our environment, as well as directing our resources to protect nature, in particular, the coral reef and natural habitats in our surrounding area. For this reason, we have made provisions based on environmentalists', local government's, and other consultants' recommendations for lump sum funds against returned earnings, set aside to help the improvement of our environmental health without hindrance to the freedom of our customers' enjoyment. We have received extremely favourable support and encouragement not only from the tourism authority, but from the industry and customers as well. We anticipate from our initiatives, working along with our development and growth, a healthy increase of our earnings.

It is my pleasure to report that our initiative also features a proper "capital charge" based on the market's long-term lending rate against the corporation's retained earnings. In so doing, it expresses the corporation's firm commitment to our shareholders' entitlement to the harvest. Last, but not least, I wish to emphasize that all these initiatives are dedicated to our corporate philosophy of serving people and nature, thus, earning others' trust through our commitment and dedication to people and the environment's health.

Public Confidence, Stock Value and Conclusion

What would happen if the public loses confidence in the corporation's ability to satisfy investors' desire for profit? If so, would the corporation have difficulty raising needed financing in the stock exchange? Under the circumstances, this need not be a serious concern because of the following reasons.

The stock exchange, the single most powerful money swapping machine ever invented by humans, is a truly invisible hand and, more often than not, determines the destiny of a corporation. As it is an invisible hand (in fact, the "invisible" cannot see nor have feelings for people and the environment), it is guided by people: stock analysts, stockbrokers and a host of others dispatching information and "producing information" for the investing public, telling them what to invest and where. What is needed for a corporation to attract investors is to provide the public with corporate performance information and let the stock analysts analyze it, to prove to the public that the corporation is worth investing in.

The entrepreneurial approach to corporate management does not encourage poor corporate performance. On the contrary, the purpose of this book is to encourage corporate managers to manage the corporation as a community of entrepreneurs to create wealth and add value. This is a long-term management strategy both for the good of the corporation (including everyone in the corporation) and society. Therefore, this caring for people and corporate environmental commitment and practice should project an excellent image, which combined with a strong entrepreneurial drive, would push the corporation to further realize its growth potential, even though it questions the traditional motherhood, "profit". In fact, there is no intention of suggesting that a corporation should not make "profit", just that the real meaning of "profit" be realized and the misconceptions of what "profit" has done to people and the environment exposed. Make "profit" by all means, but without harming the environment, nor taking away our future generations' right to live and enjoy life as we do. The circumstances should provide stock analysts with favourable input, hence, recommending the corporation to the public as an environmentally friendly and a people focused entrepreneurial organization. Why should this affect the public's confidence in the corporation's earning ability?

Someone once said: "We are the prisoners of our own technology." Most likely we will be prisoners for a long, long time, unless we free ourselves by revamping management thinking. The purpose of this book is therefore an attempt, in essence, to share the following:

1. Recognition that a corporation is not merely a legal entity prescribed by the law, but a social entity.

2. As a social entity, a corporation is not merely responsible to its stockholders, but to all contributors to the corporation.

3. Understanding that capitalism is an economic process, that only facilitates capital accumulation. Entrepreneurism, on the other hand, provides opportunities to individuals to acquire ownership and to create and innovate for the individual, the corporation and the common good.

4. Entrepreneurship is not merely about starting up a new venture associated only with business undertakings, but a wealth-creation and value-adding process for the individual, entity and society.

5. Realization that the purpose of a corporation is not merely as a money-making machine, but to serve people and earn their trust through providing goods and services to the consumers' needs and at the same time caring for people and the environment. Nevertheless, we can all realize our expectations, including making a "profit".

6. If individuals are to work in the organization as entrepreneurs, the corporate goal should not be determined from the top and handed down. It must be a collective effort, starting from the bottom up and integrating individuals' personal values and culture.

7. Acceptance of the reality that currently what businesses do is merely to make different combinations from the resources available to us, thus increasing their utility function to satisfy the needs for living. The role of business managers is to allocate sensibly and use resources for this purpose, no more and no less.

8. If the corporation needs to make a "profit", make a profit by all means, but not at the cost of people and the environment.

9. Awareness of the merit of stakeholdership, a practice that provides the opportunity for employees to be able to identify themselves with the corporation.

10. Development of a system for environmental accounting, through which a corporation can make provision for the central thought of care for environment with substance.

Concluding Thoughts

There was a time when people built churches, and not only did they contribute money and material but they also conscientiously donated their own labour. The churches were taller than any other buildings, each one with a steeple and a cross reaching the sky, symbolizing a human need to be closer to God and to seek his blessings for peace, harmony, love and brotherhood. Today, we witness the erection of corporate skyscrapers and office towers housing banks and other financial institutions. These buildings not only surpass church steeples with their crosses, but appear to be competing among themselves to be the tallest of them all. They too are symbols, but not necessarily symbolizing or reflecting human needs, but demonstrating corporate power. Perhaps corporate and bank executives simply wish to tell the world that with the blessing of capitalism, the rapid accumulation of corporate financial wealth, combined with their strength they will conquer the world. The question is, is this what we want?

Human beings have come a long way to reach this stage, evolving from four-legged animals or, if you prefer, from the day our first parents, Adam and Eve, committed their mortal sin and were driven out of the Garden of Eden to become sophisticated creatures on earth. Using our intelligence, physical and mental abilities we made it possible for a lot of us to live in style, with some rich enough to be able to command sufficient resources to provide for all the needs of the people of a small nation.

Picking no bones about these attainments, the high standard of living for a lot of us, to a large extent, is the result of corporate managerial efforts, but it is at the cost of taxing labour residuals, otherwise causing job loss and displacement, excessively tapping resources from the future, and endangering the health of the natural environment; in the name of profit to compete in the global market and earn a sufficient rate of return for investors. Even after record earnings amounting to hundreds of millions and billions of dollars, some corporations such as General Motors of Canada, and British Airways, are still attempting to increase their earnings at the cost of job losses to those who made the "record earnings" for the corporations possible. These trends and ill practices should not continue, because they are the cause of human misery, social unrest, poverty, crimes, environmental disasters, a drain on resources, and above all, endangering the human life-support system, thus limiting the possibility for the continuation of life through our children and children's children.

Capitalism is not the solution to this misery, as it is the cause, and it only drives people apart and further damages the environment and the life-support system. Capitalism is built on the assumption that resources are plentiful, people can be replaced as if they were disposable materials, rubbish can be dumped on other countries which are in need of money, and environmental support is always present. But we know only too well that this is not the case.

What could the solution be? Solutions do not come from the government, politicians, academics and professionals – although the government could provide a favourable environment to facilitate freer flow of resources within the political boundary; academics could venture into advanced thinking and provide guidance (please note that it is an insult to learners' intelligence and a tragedy to academic advancement if academics rely on case studies and empirical testing, because these are only research into performance of the past, not development. Consequently, it would not be possible for them to reach the frontier and break the knowledge barriers to provide forward thinking which society and business need so badly for guidance); and professionals could provide meaningful and useful tools, particularly accountants and economists who could provide a more meaningful method of cost measurement, and the calculation of GDP and GNP without misleading the public. Nonetheless, it is only the corporate executives and managers who could act and turn the trend around to build a healthy economic future.

There are sixteen chapters in this book all with one single purpose. The purpose, to put it simply, is to encourage corporate managers, individually and collectively, to change their mindset and get away from the single narrowly focused profit objective and become broader-minded individuals of corporations, managing our limited resources in the interest of themselves, society and humanity as a whole. This is entrepreneurship, and corporate managers are true entrepreneurial managers.

With the permission of a reviewer of this book, I would like to borrow his comment to reflect what is on my mind:

> We all talk about things like values, social responsibility, ethics, employee motivation, future driven management, etc. This book provides the vehicle to pull this all together in a methodology that can work and truly make the firm a means to benefit overall society, and itself.

As we enter a new era of economic cooperation and freer trade to facilitate resource flow for people, it is the corporate managers who could make these dreams a reality. Corporate managers need to realize that a corporation is a public institution, not just a money-making dream machine for a few. Certainly, corporate power should not surpass the church, nor be above religion, or tell the world that they are here to conquer instead of serve. A corporation must be a community of entrepreneurs, managed by the people, of the people and for the people and the surrounding environment with roots developed for the future.

Life is a one-way trip, no one has a return ticket; therefore, the purpose of life is life itself, not merely "life" for the individuals present, but the continuation of life through our children and children's children. If corporate managers do not care about people and the future, but are only interested in making "profit" for a few, what is the purpose of life?

I would like to borrow from the wisdom of an underground fighter in the Second World War who left a note while in captivity before his execution (though these are not his exact words, the meaning is clear, it is recorded in a park situated in Lake Como, Italy), it reads: Like the falling leaf, I am just about to drop to the ground where I can fertilize the soil for the future....

Questions for discussion

Introduction

The following is a set of 160 questions. None of these questions requires an absolute or correct answer; in fact, there may not be a correct answer, but through discussion and exploring the wisdom imbedded in these questions, we may have better answers.

Bear in mind that some people with strong views might believe both the book and questions to be wrong. Then why devote so much effort to write this book and ask so many questions which may be deemed wrong? The answer is simple: if I am allowed to be vocal, the "wrong" itself comes with birth; remember the story of Adam and Eve, the reason why we are here on earth. Moreover, in democratic countries such as the US, Canada and the UK, it is claimed that we have the right to be wrong, and it is a constitutional right to be defended that we are not wrong.

If you appreciate the above, you would agree that having the best brain in comparison to any other living creature, we must think, and think more rather than just follow others. Be a follower, by all means, but do think. To think is the whole purpose of this book, and hopefully it will make people think. Do not be indecisive; learn to think with your feet planted firmly on the ground.

Chapter 1

1. Should corporations pay "environment tax"? At progressive rates? Why?
2. According to what you know about "entrepreneurship", is it possible to set a corporate goal so as to obtain the desirable (or maximizing) "rate of return on investment" for shareholders, and at the same time make a corporation a community of entrepreneurs? Of course, first, you must define who an entrepreneur is .

3. Comment on: Union Carbide's "corporate decision" to liquidate a substantial portion of its assets and give them out to shareholders in special dividends after the Bhopal accident in 1984 that killed more than 2,500 people and seriously injured approximately 100,000.

4. What is corporate value and how should it be managed?

5. Comment on the author's definitions of "entrepreneurship" and "entrepreneur".

6. What is an "entrepreneurial corporation"? Is it possible to build an entrepreneurial corporation on the basis of the author's definition? Why?

7. Who governs a corporation?

8. Discuss: "A great business is really too big to be human" (Henry Ford).

9. How do you make a distinction between political democracy and economic democracy? Are they mutually exclusive, or interdependent? Can people have political democracy before economic democracy?

10. What do you perceive "dynamism in capitalism" is?

Chapter 2

1. Is there any difference between "making money" and generating "profit"? If so, what are the differences? If no why not?

2. The fisherman had a good catch from the sea and there is "profit" for the fisherman, but whose loss is it? Discuss.

3. "If the capital of a business is impaired, there is no profit." Therefore, we could only have "profit", if the capital of a business remains intact. Explore and discuss with your own understanding "capital", and how we determine "capital"? If we cannot determine "capital", how is it possible to have "profit"?

4. "If management is lagging behind technology, there is no profit." Do you agree? Why?

5. Should a corporate manager be aware of accounting theory developed over a long period of time for the purpose of measuring "cost" and "profit"? Is this simply a job for accountants, or the concern of corporate managers? Why?

6. One of the accounting standards governing reporting accounting information is the freedom from bias. Under the circumstances, why does accounting recognize only the cost of excavating the ground, the digging, and the mining of underground resources, but nothing about adequate costing for the resources themselves? (Note: depletion is a simple method to convert resources into cash, not the cost for underground resources.)

7. Economists have attempted to account for environmental cost from the basis of ecological and environmental damages. On the other hand, attempts and efforts made by economists have hardly made any impact on corporate managers' decision in respect to environmental cost. Do we have the need for such costing? If yes, what should we do? If not, why not? Is there any accounting theory that eventually will lead to accounting recognition of environmental cost? If so, how should it be developed? Use your own imagination and state your opinion, and do not worry if others feel you are wrong, because you have the right to be wrong and the right to be defended that you are not "wrong".

8. What is the purpose of "profit" determination? (Do not simply dismiss this as an obvious question, but try to give an answer. Perhaps it might not be so obvious.)

9. Cost means sacrifices. How does accounting measure sacrifices?

10. "Profit is the cause of poverty, but poverty is not the reason for people wanting to make 'profit'." Discuss. By the way, do we really know what "profit" is?

Chapter 3

1. "It is the goal of making profit for the firm that prevents new thinking on the need for business management to break the frontier barriers of knowledge." Discuss.

2. To your knowledge, what is wrong with our corporate management? Explain in full. If you feel there is nothing wrong with the way we manage our corporations, please give your reasons.

3. What is quality management? What is quality management and beyond? Scrutinize it carefully, and identify how much difference there is between the new idea of "customer-driven" and "customer-oriented marketing strategy"? Then discuss whether

there are any new distinct elements of "system thinking" claimed in "quality management".

4. Explore some corporate managers' challenges for the brand new century.

5. What is "management by fear"? What is the purpose of creating a "fear environment" in the process of managing a corporation?

6. Between mastering the computer and managing the people in an organization, which would be your choice? Why? Considering all things are equal, which would you prefer, a clean humanless environment or an environment with human interactions?

7. Do you perceive that management is nothing more than solving problems? On the other hand, there are others who consider management's function is to allocate and manage resources for specific purposes. Solving problems is part of the management process, not the purpose itself. Moreover, problems in management are really a matter of perception. Some people view problems as challenges and others perceive them as opportunities. Comment.

8. Would you consider "entrepreneurism" is a sensible alternative to capitalism on one hand, and socialism on the other? Why?

9. What is intrepreneurship, intrapreneurship and interpreneurship? Try your luck, you may not even find these words in the dictionary. (Note: Intrepreneurship: home grown within the corporation. Intrapreneurship: spin-off from a corporation. Interpreneurship: in a family business environment.) Find out the differences between the three, how do they relate to "entrepreneurship"?

Chapter 4

1. Why is "management" unable to be a distinct profession, like engineering professional accounting, etc.?

2. Is it appropriate to say: to manage a business professionally, but not to make an individual a professional manager? Perhaps you may feel all managers should be professionalized? Why?

3. Business managers have the noble function of allocating scarce resources in the interest of creating wealth and value. Why do business managers seem to set their mind only on increasing revenue and profit? Explain fully.

4. The author perceives that there are at least five plus one fundamental business educational objectives, including:

   (a) To develop an individual to be a good citizen.

   (b) To develop an individual to meet the societal needs.

   (c) To develop an individual to have a "profession", or the ability to earn an income in the society.

   (d) To develop an individual's capacity to develop his or her innate potential.

   (e) To develop an individual's ability to create wealth for him or herself and add value to society.

   + the one

   (f) To teach individuals how to make money for the corporations.

   Comment on the above. You may wish to add or delete any one or all of them from the list. At the same time, examine whether they are verifiable, with reference to what you have learned from "B" school, and comment.

5. Can we have a general theory of management, if so, what would the theory be? What is the purpose of having such a theory?

6. There is a thought that in order to manage business, we need information, because with sufficient information, we will be able to make a business decision and act on the decision we made. Today, the development of Internet is perhaps the result of pressure from management needs that made it happen. In the same view, it is the invisible hands of market economy that prompted the engineering of the dynamism. Comment.

7. The American economist Milton Friedman advocated that to take on economic, social goals would violate the interests of the stockholders, resulting in business imposing its views on others, and would be fundamentally subversive to the free enterprise economy. Comment.

8. What is corporate responsibility? What is social responsibility? Who should assume social responsibility; the government, the church, the United Way (annual charity drive), or is it not the responsibility of corporations (the employers of working individuals)?

Chapter 5

1. Is capitalism a political catchword? Do you agree? Why?

2. Comment on: The environment has been the silent victim of "capitalistic growth".

3. Connotation of "capitalism" has evolved with time. Capitalism in the early colonial days is quite different from capitalism today. Accordingly, the single largest investors in corporate equity are pension fund trustees. In another words, today, capitalists are not only those who are rich, but the working class who invest the fruit of their labour into the ownership of large corporations. Under the circumstances, corporate managers are, in fact, working for wage earners on both accounts: labour of yesterday and today. This could work beautifully, if there is always "full employment". On the other hand, if there are two digits rate of unemployment, where is the unemployed labour's money to invest in pension funds? Discuss.

4. Capitalism is based on the idea of market economy which magically claims to be regulated by the unbiased mechanism of the invisible hands: the pricing system based on supply and demand. On the other hand, the unbiased mechanism of the invisible hand has no feeling for people and the environment. Supposedly, for the satisfaction of capitalism it advocates that there shall be no government intervention, under the circumstances, does the mechanism have any built-in system to have regards for the environment and the life-support system for our future generations? If we do not care to adjust the irregularities will they be the cause of unemployment, human miseries and deterioration of our life-support system? Can we find that mechanism? The answer is "yes": to have a more accurate system. Do you agree?

5. What do GDP and GNP measure, if they do not take into consideration the cost of the environment and the rapid depletion of natural resources? No one has seriously considered a change.

6. Think and think hard, how can corporations be enterprising in the marketplace and in partnership with the environment?

7. "Poverty is not the enemy of the environment, blindly expanding capitalism is. In fact, people who have less will make everything reusable, but it is the rich who throw everything away as waste, even though they are very much useable." Do you agree? Why? Please bear in mind that wanting everyone to care for the

environment and sensibly use resources (use things reusable), does not mean everyone should be poor.

8. Recycling is better than making everything disposable, but recycling requires a reproduction process and creates waste. The best way is to reuse everything that is reusable. Why then did Coca-Cola and other soft drink companies practising the "reusable" strategy, change to the use of recycled containers? The milk processors are even worse, they are using throwaway cartons instead of reusable milk containers?

9. Try to make a comparison between what capitalism contributes and what it destroys. And think really hard and see if we have an alternative to better serve humanity all around (including our future generations).

10. How can environmental conscious corporations at home compete with those in some other nations where environmental care is not (less) an issue? Some corporate executives have apparently made a decision: "Let's move our manufacturing outlets to another country." Comment.

Chapter 6

1. What do you know about capitalism? By depositing money in the bank would it automatically make you a capitalist?

2. How does capitalism relate to the market economy? Why should a corporate manager know everything about "capitalism"?

3. The downfall of the Berlin Wall and disintegration of the "Soviet Union" have been noted as triumphs of "capitalism". Do you agree? Why?

4. The fundamental of entrepreneurism is built on "wealth" and "value". Drug traffickers, crime organizers, and industrial polluters causing damage to the environment and people, are endeavouring to make money for themselves, but do not add value to society, therefore, they are not entrepreneurs. Do you agree?

5. Entrepreneurship is not about telling people how to make money through a business endeavour, but to create wealth and add value for the individual and common good. Do you agree?

6. According to Lipson: the fundamental feature of capitalism is the wage-system under which the worker has no right of owner-

ship in the goods which he manufactures. He sells not the fruits of his labour but the labour itself. Do you agree?

7. How do entrepreneurs differ from investors?

8. Are land speculators entrepreneurs, investors or capitalists?

9. One of the author's MBA students commented on the author's definition of entrepreneurship, as being prescriptive. He similarly noted, that socialism is also prescriptive, and it failed because it does not reflect reality. Hence, he said: "Professor Kao's definition would have a hard time convincing others because it does not reflect reality." The author responded to him: "The Ten Commandments can also be described as 'prescriptive', but they are almost the fundamental laws which the legal system of the western world is based on." Do you agree? Why?

10. What is the role of the government in the "entrepreneurial" society?

Chapter 7

1. Do you agree with Professor Das' estimate of the value of a 50-year-old tree? Why? If you do not agree with his claim, how much would you value a 50-year-old tree at?

2. "Vision" is a big word for the average corporate managers, particularly, if they are pressured by short-term performance measurement, mainly, to obtain the required ROI for the corporation. Under the circumstances, how could we convince an individual to commit and develop his or her own vision about the future which could be so remote from reality?

3. What is a corporate vision, how does it differ from the vision of an individual?

4. Can the individual's vision be aggregated within an organization?

5. Comment:

   (a) A customer-driven vision that specifies the following:

   • We will be recognized by our customer as producer of the highest quality _____ in the world.

   • We will be seen by our customers as the suppliers of choice of all our products.

(b) A profit-driven vision that specifies the following:

- Our reason is to make money for our stockholders.

- Our ultimate objective is to maximize profit.

- We still build a revolutionary new _____.

6. Look at what is happening in the world. As a corporate manager, should you be concerned about tensions between North Korea and South Korea? China–US relations? ASEAN's regional developmental strategy? European Union's unifying currency (Euro dollar) development? Among other things, on the one hand, invisible trade barriers exist in some countries which claim to liberalize trade, but put up barriers on the other?

7. How do you relate unifying theory with corporate management?

8. Try to outline what physics, chemistry, and particularly, bio-chemistry have contributed to business management?

9. No one in this world could solve the damage we have done to our environment and life-support system for our future generations, because what we see, feel and experience are symptoms of the greatest problem of mankind. Outline what you feel corporate managers could do to help, thus, to lessen the damage to our environment.

10. There are two contrary views in respect to the allocation of our natural resources: To safeguard limited resources for the nation's own use, and the other to simply open the market and let the invisible hands do the work, so there will be more development opportunities to create more millionaires. As corporate managers, how would you deal with the directing and allocating resources responsibility?

Chapter 8

1. In your opinion, should corporations have their own philosophy, or should a corporation's philosophy be one of money making?

2. The breast implant lawsuit is costing Dow Corning not just money, but its corporate image as well. Was Corning's management decision to become involved in the breast implant business prompted by "customer satisfaction is the only objective to ensure stable and continuous business increase"? Was it wrong for Dow Corning to enter the breast implanting business? After all, it was for customers' satisfaction!

3. What is the difference between added value and profit? Discuss.

4. People tend to care more about things they own, but not the things others own or which are owned by the public. On the other hand, ownership is merely a relative word, in some way a state of mind. In a nutshell, so long as an individual is able to make a decision over matters (or resources under his or her control), in effect, he or she is the owner of the matter or resources. Why is it then so important to acquire the status of ownership and push it to the extreme? More often than not, it initiates conflict and entails making great sacrifices.

5. Ownership is all about making decisions to allocate resources (including human resources) for predetermined purpose(s). Do you agree, why?

6. In a corporate structure, a minority shareholder has very little influence over matters affecting the corporate well-being. How can a staff shareholder, with a very small fraction of equity, exercise his or her proprietary right to influence the making of a corporate long-term entrepreneurial development strategy?

7. Once a corporation decides to implement its stakeholder ownership participation, why should the company need to plan for financing?

8. Staff participation in corporate equity and management are things of common interest in the academic community and corporate environment. In Europe, share participation is not unusual, and union representatives participating in management committees can be found in a large number of US and Canadian corporations. Why then, when labour-management problems occur, more often than not, cannot be resolved, it ends up with the unions taking action (such as strikes) against the company? Is it because there are corporations with no such schemes, or the scheme failed to function?

9. Is it desirable for a corporation to take care of its employees from cradle to grave?

10. Assume a corporation's management dedication to serve people is its philosophy and makes no specific reference to profit-making in the interest of its shareholders, under the circumstances, will investors still be interested in investing in this particular corporation?

Chapter 9

1. Value is a personal property, it varies from individual to individual. On the other hand, a corporation is an entity made up by humans (even in a highly computerized corporation, it still needs at least one person to operate the computer), but how is it possible to integrate an individual's value as an aggregate, so it can be used to guide an individual's behaviour and action?

2. Suggest a corporate value which every individual in the organization can identify with.

3. What is the role of corporate managers in respect to the development of a corporate value that includes personal value of every individual in the organization?

4. How much is value worth? What has value got to do with making money? Discuss.

5. For better or worse, the United Nations has survived for more than 50 years. What are the UN's values that make this organization work in this troubled world?

6. Think of the Japanese value system. In your opinion, is the Japanese value system such as "unquestioning devotion to the national unity" the reason (at least one of the reasons) that substantiates Japan's economic success? Why?

7. Christian values have been a treasure to Westerners. Which Christian values are reflected in business practices?

8. What are the perceived differences in values between the East and the West?

9. During the depressed period of the steel industry, British Steel allocated funds to work with local employment agencies to train redundant employees and local unemployed individuals to create their own businesses, and called the programme "Putting something back". The idea was quite simple, the company felt it would be appropriate to recognize the employees' contribution and share good times as well as bad. Consequently the undertaking drew world attention. The same was attempted in Canada, but did not materialize as in the UK. Why?

10. Refer to the fish and water story (p. 175). If the staff (water) is so important to the boss (fish), why is the staff treated as disposable useless materials sometimes?

Chapter 10

1. What is enterprising culture? Why is it important to the corporation?

2. Based on the response to the above, suggest how to develop such a culture in a corporate environment.

3. The author was a Tasman Fellow of the University of Canterbury, New Zealand (sponsored by the Tasman Company), and among other things, his involvement included lecturing on entrepreneurship. One day three students came to the office and told the author: "Professor Kao, we are sorry for not handing in our project due last Friday. You can give us a zero." The author asked them why? They said: "We played football." The incident reminded the author about the countless times his students failed to meet deadlines during his years of teaching. In most cases, they gave excuses such as: "I was sick", or "my grandmother died last week" (even though the deadline was set two months earlier); and requested for late submission. In the author's thirty years of teaching experience, every single student pleading for late submission never admitted: "I couldn't hand in the assignment, I deserve to have a zero." Why?

4. Why should there be any managerial cultural differences between entrepreneurial managers and corporate managers? If you feel there should not be any differences between the two groups, explain why.

5. Why do some sub-cultures exist in your organization? Are these sub-cultures a hindrance to the development of entrepreneurship? If so, how do you suggest overcoming these hurdles? What are the roles of corporate managers under the circumstances?

6. How do you develop a positive "yes" sub-culture, rather than "no" or "but" or "we can never make it" sub-culture?

7. Staff control is a heavy function in large corporations. In your opinion, is staff control function a blessing or hindrance to progress? Explain. If you perceive staff control as a hindrance to the corporation's enterprising culture, what should you do to "turn this problem into opportunities"?

8. Can you make staff functions more flexible to reflect the dynamism needed for development and growth?

9. To distrust others is a common behaviour in any organization, although in Western culture, we are supposed to observe the

very fundamental value of believing that "a person is not guilty until proven guilty". On the other hand, by overly trusting people, a person can be "burnt" over and over. Borrow John Bulloch's (president of Canadian Federation of Independent Business) description of an entrepreneur: an entrepreneur is: Total scar tissue/Total body area = 1. Trusting people too much would result in a modified version of John's description: Total burnt marks/Total body area = 1. The question is: how do we develop a culture to make people trust people when they already have had so many bad experiences?

10. "I am the boss!" If this is a nameplate placed on the door and written all over the face of your "boss", what would you do? Tell him or her there is no need to do that? But he or she has the right by appointment and the right to tell you: "I am the boss!"

Chapter 11

1. To your knowledge, how many major human conflicts in history (go as far back as you can) were caused by ownership struggle; how many were caused by other reasons?

2. Why does a country need immigration and customs controls?

3. Other than war, any matters relating to ownership are negotiable, the only difference is price. Do you agree? Why?

4. The stakeholder idea is not new, there has been sufficient claim that multinational corporations are making themselves good citizens of every country and every community. Britain's efficient retail industry has demonstrated how companies can work together for their "common good", just as in countries such as Brunei, which are virtually dependent on a single company. It is the idea of "common good" that the stakeholder concept is advanced. As a matter of interest, find out how many companies in your neighbourhood actually consider the community, and the corporations' staff, to have a stake in the companies.

5. Comment on: The stakeholder idea is a grassroot approach which together with the staff participation scheme, form the foundation of a strategy to turn a corporation from merely a money-making machine to a community of entrepreneurs.

6. How do you differentiate a stakeholder participation plan from a scheme inviting the staff to participate in a company's shareholdings?

7. Use Table 11.1 (page 224) as a reference. Argue for and against the use of social entity approach to form corporate management strategy.

8. Why are some things extremely important to us not part of the market economy, yet for one reason or another we brought them into the market jungle, things such as drinking water, scenic views, beaches and the natural environment. Based on this rate of progression, it will not be long before someone brings "fresh air" into the market economy. Based on your understanding, what caused all these? Obviously, it is the game of private owner-ship. Would it also be the cost of capitalism? Discuss.

9. The author suggests that corporate culture should be built on ownership participation (including stakeholder ownership and equity participation ownership), and delegating decision-making rights. Do you agree? Why?

10. Carefully examine the suggested stipulations outlined on pages 229–30, comment on the idea and make suggestions for improvement.

Chapter 12

1. Can entrepreneurship be taught? If entrepreneurship cannot be taught, why does virtually every business curriculum in colleges and universities offer a course or programme in entrepreneur-ship?

2. Is it possible to develop entrepreneurship whereas the corpora-tion's goal is to make money and satisfy shareholders' desire for the rate of return on their investment?

3. People's mindset can change. What external forces or internal drives will make a person change his or her mindset?

4. There are literally hundreds of entrepreneurial attributes identi-fied by researchers. Use the information provided in this book (or other sources), and identify those attributes applicable to a corporate environment, and discuss how to develop them.

5. Risk-taking is considered to be a single well-cited, identified attribute relating to the entrepreneurial drive development. What if a person's mindset firmly believes in the idea of "no risk, no gain, no risk, no loss", and just absolutely refuses to take any risk? What can be done to change his or her mindset?

6. Japan's economic success, in all due respect, has been its determination, not just individually but nationwide, to break the barriers of knowledge and technology. While other countries are aiming to correct their wrongdoing, Japan has gone its way by re-engineering the structure. The improvement of the car industry's efficiency performance and the advancement in the field of electronic technology are examples in point. Even though it was Silicon Valley and Bill Gates of the US who made the headlines, the Japanese learnt from the best and surpassed them. Today in Japan, why do businesses, small or large, confront difficulties to stimulate entrepreneurial drive within the organization? Feel free to think, express yourself and discuss what magic could make "entrepreneurship" tick in the corporate environment.

7. Some supportive attributes which may not be so popular as risk-taking, flexibility, and leadership, work well for a large number of people, including the author and his students. It is the illegitimate attribute of "don't put people in a defensive position". Try it yourself, and find out if it works.

8. How important is it to be aware of and use the interrelationship between various attributes in managing people and resources. Explain. If you agree with what is shown in Figure 12.2 (page 255), would you like to try it and see how it works?

9. Tolerance of failure looks good on paper, and so does lip service, but in reality, if the person responsible fails the task and gets away with murder, it will set a bad example, and lower work standards. Moreover, a great deal of injustice is done to those who worked hard and succeeded. Do you agree? Discuss.

10. A computer program is a reflection of the intelligence of the programmer. Can the computer help to develop entrepreneurial attributes? That is, if humanity is still useful, overall the strategy is to develop entrepreneurial attributes.

Chapter 13

1. To your knowledge, does organization facilitate communication? On the other hand, it could be a hindrance to communication. Comment on the statement.

2. In our high-tech society, the plumbing trade is highly computerized with the aid of the computer. Would it be possible to

employ a computer to treat corporate organizational illnesses? For example, people in the organization have no entrepreneurial drive.

3. Bureaucratic organization pyramids emphasize the importance of rank, position, parking space, key to the executive washrooms, and above all, the chain of command. How does this type of organizational structure fit in with an entrepreneurial way of managing? An entrepreneurial way of managing may be described as highly personalized, tends to focus on people interactions, and decentralized decision-making, and everyone in the corporation has a stake in it.

4. Profit centre practice is designed to give division managers autonomy. Why is it possible to stimulate division managers to concentrate on the long-term benefits of the corporation (division)? Better still, under the profit centre structure, is it possible for the division managers to put long-term benefits before short-term divisional ROI performance?

5. Internet has surpassed e-mail, turning out to be the communication wonder of the twentieth century. Will it help communication within the corporation? Communication means more than just the information flow.

6. How does the spin-off organization grow up to be an "adult", rather than a "baby adult" of the parent corporation?

7. There is no such thing as an ideal or perfect organization, but there could be better organizations. According to what you know about organization, which organization would be most appropriate for a corporation that advocates entrepreneurship? Why?

8. Can staff functions create wealth and add value to society? Explain.

9. What should be the role of staff in an entrepreneurial organization?

10. Would computerization reduce staff function needs? With more and improved computerized accounting reporting and analysis through Internet, would it be possible to eliminate the controller's function in an entrepreneurial organization? Discuss.

Chapter 14

1. Years ago, there was a film entitled *Five*, which described the aftermath of a nuclear war. The whole world was left with only

five people including one woman. Although they were not supposed to enter a city because of the high level of radiation there, one of them disregarded the danger and went. While there, he grabbed all he could; money, gold, jewellery and some food but later he realised that these things were worthless to him. There was nobody to share them and he later died from radiation before he could go back to the other four. Then a fight broke out between two of the men over the woman; one man was killed and shortly after another died of old age. Now remains only a man and woman – the new Adam and Eve started another round of human society. The moral of the film: it is simply all about people. Here's a thought-provoking question: How do you think an entrepreneurial corporation should be managed?

2. How can Japanese companies keep jobs before profit, but some other corporations get rid of people (in the name of changing cost structure) for reasons of improving profit?

3. There are 26 cultural values listed on pages 289–90. Look them over carefully, and see how to apply them to shape management practice in an entrepreneurial corporation.

4. Suggest any model that will make managers and individuals work together in order to cultivate and develop individuals' need for creativity and change.

5. A corporation is not human, but could have a culture of its own. For example, fair business dealing is a corporate culture, like some Hollywood film industries that put social issues before the box office. Try to identify corporate culture known to you and discuss how the culture affects individuals in the firm.

6. Consumers are not kings and queens, but ordinary people just like you and me, our children, and their children and future children. Do you agree? Why then, in corporate planning, some, if not all, corporations made no attempt to think about the needs of the future?

7. List some of the reward systems you know, other than money. Discuss these systems in relation to developing an individual's entrepreneurial drive in the organization.

8. Comment and discuss remarks made (quoted in Chapter 14, pages 297–98) by Mr Wee Cho Yaw with direct reference to:
   • Short-term gains versus long-term benefit
   • Professional CEOs' professionalism

9. The better punishment is letting the individual who did wrong to come to self-realization. Do you agree? Why? If so, what should be done to make individuals realize their wrongdoing?

10. Discuss the manager's responsibility in respect to entrepreneurial skills development for the individual.

Chapter 15

1. Is it possible to use the bird and the cage story as a motivational strategy for entrepreneurship development? Why?

2. "Poverty itself pollutes the environment." Do you agree? Why? Compare "poverty pollutes the environment", aluminium foundries pollute the environment, chemical products pollute the environment, and the people who are saying all these things pollute the environment. Why did the World Commission on Environment and Development single out only "poverty"? Poor people may chop down trees to support themselves or catch fish for food. But how about the car manufacturers, paper mills, tanning industries and all those industries discharging chemical waste into rivers, lakes and oceans, and last but not least, nuclear testing? Discuss.

3. List some sustainable development ideas.

4. What is the responsibility of accountants in respect to concerns for our environment?

5. Think of some measures for environmental care that a corporation could observe, and some environmental friendly products that a company could consider.

6. What do you consider to be a workable "environment accounting"?

7. Do you agree that polluters must pay for the environmental damage? How should such damages be measured and cost be determined?

8. A company's production process and waste disposal, conceives environmental damage potentials but could not possibly make adjustments but continue with the process. In your opinion, should the management make necessary provision from its earnings for possible costs incurred in the future?

9. Discuss corporate managers' responsibility in respect to resources allocation. Would it be sufficient just to use the resources efficiently and effectively to achieve the company's goal?

10. There are five factors for us to consider:

- Rate of consumption (Cu)
- Rate of technology advancement (T)
- Rate of making new combinations (I)
- Rate of resources consumption (R)
- Rate of management skills improvement (M)

If Cu = T = I = R = M, under the circumstances, is there a temporary equilibrium? What must a corporate manager do in order to help achieve the above? On the other hand, are there other factors to consider? If so, what are they?

Chapter 16

1. How should a corporation's performance be measured? Develop a set of criterion for the measurement.

2. Make a comparison of corporate reporting under (a) the legal entity concept and (b) social entity concept.

3. What should be included in the corporate reporting in respect to the care of human resources, environmental protection, utilization of resources and the degree of innovation and inventive activities.

4. What is operational residual? What is the purpose of such accounting?

5. What is the difference between the use of "financial income", "operational residual" and "value added" as a corporation's performance indicator?

6. As a stock analyst, how do you disclose corporate information to potential readers and corporate investors in respect to a corporation's responsibility to environmental care and resources utilization?

7. Refer to Chapter 2, based on the accounting theory known to you, what part of the theory that suggests environmental care cost should be part of the cost of doing business?

8. If you are a member sitting on the Accounting Standards Board, give your review for the use of the value-added approach instead of the "profit" approach for corporation performance determination.

9. What must we do in order for accountants to recognize environmental protection cost as a legitimate cost?

10. Develop a new formula to account for a nation's GDP and GNP.

Notes

Chapter 1

1. It may not be a well-known fact that in some factories in China the factory director or manager's position is an elected position. Accordingly, any member of the factory could enter into competition for the position, much the same as in a political campaign. A candidate can state his or her policy for managing the factory, and all other members of the factory will exercise their democratic right to vote on the best person for the office.

2. The best competition can be simplified by a remark by an American clergyman in his Sunday (7 July 1996) sermon. He said: "Somewhere in the United States, a street has four donut shops. The first one has a big sign which reads: the best donut shop in town. The second one puts up a bigger sign that says: We are the best donut shop in the United States of America; the third one has the biggest sign with very bold letters: We are the best donut shop in the world; the fourth, a small shop also has a sign which says: We are the best donut shop in this street."

Chapter 4

1. Car manufacturers always recall cars and trucks for different reasons. There were at least two incidents involving two different car makers whose executives decided not to recall cars or trucks with design flaws (one involving a potential fuel tank explosion [Ford Pinto], and the other a crash safety problem [GM Trucks]).

2. A typical problem-solving approach can be seen in *A Statement of Basic Accounting Theory*, published by the American Accounting Association, 1966, p. 50.

3. The following statement made by Rawl Lawrence, Exxon Chairman, reflects at least one corporate executive's mindset: "Americans could identify far more easily with the wives of fishermen weeping for the loss of livelihood and family savings than with Rawl's boast, 'I am confident that Exxon's tractional financial

strength will not be impaired by this major accident'" (in Hawken, 1993, p. 118).

Chapter 5

1. Other than China, Cuba is perhaps the last major socialist state. However, it is moderately shifting its communist ideology towards José Marti. [Marti was a national hero who died in battle against the Spaniards about 100 years ago (1895).] Small entrepreneurs are welcome in Cuba in recognition of their role in creating wealth for themselves and adding value to Cuba's economy.

   An article entitled "Cuba Concentrates on Hedonism", written by Hugh O'Shaughnessy (*Calgary Herald*, 5 April 1995), describes Cuba's recent effort at economic recovery through tourism while still engaging in Cold War with the US.

2. *Costa Rica Today*, 16 March 1995, p. 26, in a paid advertisement of the Costa Rican Tourist Board.

Chapter 6

1. Entrepreneurism is an ideology based on the individual's need to create and/or innovate, and transform creativity and innovative desire into wealth creation and value-adding undertakings – for the individual's benefit and common good.

Chapter 7

1. The military cold (at times even hot) war between the old nationalists and communists was an on-going struggle even before the war ended.

Chapter 9

1. For more of the idea of human economy, see Halal (1986, pp. 114–16).

2. See Figure 7.3 in Kao (1995, p. 220). It reads:

   If a lily pond contains a single leaf on the first day and every day the number of leaves doubles, when is the pond half full, if it is completely full in thirty days? (Source: Foreword in Clifton and Turner, 1990.)

3. Murayama (1982). There was also a news item reported in a US newspaper (unfortunately, the author is unable to identify the source) which reflected how Japanese view their cultural importance. The story was about a Japanese child who had returned to Japan after receiving a number of years of education under the US system. Upon returning to Japan, the child continued his studies in a Japanese school. One day, the parents of the child received a letter from the school asking them to tell the child not to ask so many questions because his behaviour interrupted class.

4. It was in the author's note made from Hicks' Nobel prize inauguration speech. These may not be the exact words, the author assumes the responsibility for their accuracy.

Chapter 11

1. Cold peace or hot war; German Chancellor Helmut Kohl made a donation of three billion marks (US$1.4 billion) to Russian President Boris Yeltsin's re-election campaign (*The Straits Times*, Singapore, 25 March 1996).

2. A typical hardcore traditional view about a corporation is one advocated by Anthony Dearden and Vancil. It is old, 1972, from the traditional camp, but no one has yet topped it. According to them: "A business is owned by its shareholders. They have invested money in the business, and they expect a return on this investment. Profit is the measure of this return. Presumably, the higher the profit, the better off are the shareholders" (p. 93).

Chapter 12

1. This can also be changed, if a person is jobless, or starving, like in the "60 Minutes" TV programme (8 April 1996) shown in the US. Corporate downsizing put countless mid-managers and corporate vice presidents out of a job, out of a government programme and they became "persons who do not exist in the system, because they have been removed from the computer list". Surely, even though they are not traditionally considered to be an "entrepreneur", even they do not fit the "owning your own business entrepreneurial type", they may be forced to start up

their businesses just for reasons of surviving and keeping their dignity which means, their mindset can be changed even though "entrepreneurship" cannot be taught.

2. A detailed survey in the process of analyzing which is to be formalized for presentation at IV ENDEC World Conference, possibly by Tan Wee Liang and Wong Soke Yin.

3. As noted by Professor Tan Teck Meng, Dean of Nanyang Business School, Nanyang Technological University at the Third ENDEC World Conference on Entrepreneurship, 1993, Singapore.

4. This little remark was based on the author's experience: Once (it was 1981) driving in London, just about to pass through a pedestrian crossing, a man suddenly walked down onto the crossing. It was necessary to apply the brakes and stop the car. As the car stopped, a truck hit the back of the author's car, causing a little damage. As both the truck driver and the author got out of the vehicles, the truck driver yelled at the author: "Why did you stop?" The author replied, "This is a pedestrian crossing, I would have hit him, if I didn't." The truck driver said: "So what, he can be repaired!"

5. The author learnt this from Wong Chi, a lecturer at Shandong Women's College in China. She used this story to illustrate a simple wisdom of "value", while she was practising teaching and the author was the supervisor to observe her teaching "entrepreneurship" under a programme funded and organized by the United Nations Industrial Development Organization (UNIDO) in 1995.

6. From a newspaper clipping, believed to be *The Straits Times*, in the East Asia section, p. 18, sometime in February 1995; a report by the paper's Tokyo correspondent, entitled: "Japan seeks ways to nurture creativity and inventiveness".

Chapter 14

1. See Murayama (1982, pp. 107, 109) for more about Confucian influence to Japanese Business Value, in particular, the Confucian Management School (Jyokyo Chuyo Keieiha) and how Confucianism impacts the Japanese Business Value.

2. Transcribed directly from a tape recording of the speech.

Chapter 15

1. As quoted by Michael Atchia, Chief, Environmental Education and Training Unit, UNEP, in Atchia and Tropp (1995).

2. This concern was expressed by policymakers, environmentalists, academics, industrialists and organized labour. For a cross section of views, see D'Arge and Kneese (1972).

3. Frankly, the author has no idea how Dow Corning handles its account under the circumstances. The entries are made out based on "Generally Accepted" accounting practice.

Chapter 16

1. Mr Wee Cho Yaw, Chairman and CEO of United Overseas Bank Group, Singapore. For more of his speech, see *The Straits Times* (Singapore), 24 April 1996, p. 28.

2. *Ibid.*

References

Alkhafaji, A. P. (1989), *A Stakeholder Approach to Corporate Governance: Managing Dynamic Environment*, Quorum Books, New York.

American Accounting Association (1966), Committee to prepare *A Statement of Basic Accounting Theory*, American Accounting Association, Chicago.

Aoki, M. and H. K. Kim (1995), *Corporate Governance in Transitional Economies: Insider Control and the Role of Banks*, World Bank, Washington, DC.

Atchia, Michael and Shawna Tropp (eds) (1995), *Environmental Management: Issues and Solutions*, John Wiley, New York.

Babbage, Charles (1932), *On the Economy of Machinery and Manufactures*, Charles Knight, London, reprinted by Augustus M. Kelley, New York, 1963.

Berkhof, Louis (1958), *The Doctrine of the Decrees in Theology*, The Banner of Truth Trust, London.

Birch, D. (1982), *Job Creation in America*, Free Press, New York.

Blair, M. M. (1995), *Ownership and Control: Rethinking Corporate Governance for the Twenty-first Century*, Brookings Institute, Washington, DC.

Cheah, H. B. and Tony F. L. Yu (1995), "Adaption Response: Entrepreneurship and Competitiveness in the Economic Development of Hong Kong", in *Proceedings of the ENDEC Conference on Entrepreneurship*, 7–9 December 1995, Shanghai, China.

Clifton, Carr and Tom Turner (1990), *Wild by Law*, Sierra Club books, San Francisco.

Coat, E. L. (1990), "TAM at Oregon State University", *Journal of Quality and Participation*, December, pp. 90–91.

D'Arge, Ralph C. and Allen V. Kneese (1972), "Environmental Quality and International Trade", *International Organization*, Vol. 26, No. 2, pp. 419–65.

Dearden, Anthony and Vancil (1972), *Management Control Systems*, Richard D. Irwin, Homewood, IL.

Frederick, William (1963), "The Next Step in Management Science, A General Theory", *Journal of the Academy of Management*, Vol. 6, No. 3, pp. 212–19.

Friedman, Milton (1962), *Capitalism and Freedom*, University of Chicago Press, Chicago.

Fritz, Baade (1962), *The Race to the Year 2000*, Doubleday, New York.

Gibb, A. A. (1986/87), "Education for Enterprise: Training for Small Business Initiation–Some Contrasts", *Journal of Small Business and Entrepreneurship*, Vol. 4, No. 3, pp. 42–48.

Goetsch, David L. and Stanley Davis (1995), *Implementing Total Quality*, Prentice Hall, Englewood Cliffs, NJ.

Gordon, R. A. and J. E. Howell (1959), *High Education for Business*, Columbia University Press, New York.

Grady, Paul (1965), *Inventory of Generally Accepted Accounting Principles for Business Enterprises*, American Institute of Certified Accountants Inc., New York.

Halal, William E. (1985), *The Capitalism*, John Wiley, New York.

Halal, William E. (1986), *The New Capitalism*, John Wiley, New York.

Hamel, Gary and C. K. Prahalad (1994), *Competing for the Future*, Harvard Business School Press, Boston, MA.

Harrison, Brown, James Bonner and John Weir (1958), *The Next Hundred Years*, Viking Press, New York.

Hawken, Paul (1994), *The Ecology of Commerce*, Harper Business, New York.

Hayek, F. A. Von (1937), "Economics and Knowledge", *Economica*, Vol. NS4, pp. 33–34.

Hicks, J. R. (1992), "The Mainspring of Economic Growth", Nobel Memorial Lecture, in *Nobel Lectures, 1969–1980*, edited by Assar Lindbeck, World Scientific, Singapore.

Ho, Janet (1995), "Don't Stay Put, be a Handy Man", *The Straits Times*, Singapore, Life section, 8 November.

Holt, D. H. (1992), *Entrepreneurship: New Venture Creation*, Prentice Hall, Englewood Cliffs, NJ.

Hornaday, J. A. (1982), "Research about Living Entrepreneurs", in *Encyclopedia of Entrepreneurship*, edited by C. L. Kent, D. L. Sexton and K. H. Vesper, Prentice Hall, Englewood Cliffs, NJ, pp. 20–34, 36–38.

Huey, John (1992), "Managing in the Midst of Chaos", *Fortune*, 3 April.

International Society for Krishna Consciousness (1983), *The Higher Taste*, Bhaktivedanta Book Trust, Los Angeles.

Kao, J. J. (1989), *Entrepreneurship, Creativity, and Organization*, Prentice Hall, Englewood Cliffs, NJ.

Kao, Raymond W. Y. (1986), *Accounting Standards Overload: Big GAAP versus Little GAAP*, Study series 1, Accounting Standards Authority of Canada, Vancouver.

Kao, Raymond W. Y. (1989), *Entrepreneurship and Enterprise Development*, Holt, Rinehart and Winston, Toronto.

Kao, R. W. Y. (1990), "Who is an Entrepreneur?", in *New Findings and Perspectives in Entrepreneurship*, edited by R. Donckels and A. Miettinen, Gower, England.

Kao, Raymond W. Y. (1993), "Defining Entrepreneurship: Past, Present and ?", *Creativity and Innovation Management*, Vol. 2, No. 1.

Kao, Raymond W. Y. (1995), *Entrepreneurship: A Wealth-Creation and Value-Adding Process*, Prentice Hall, Singapore.

Kendall, N. (1994), *Good Corporate Governance: An Aid to Growth for the Smaller Company*, Accountancy Books, Central Milton Keynes.

Keynes, John Maynard (1964), *The Theory of Employment Interest and Money*, Macmillan, London.

Kirby, D. A. and Y. Fan (1995), "Chinese Cultural Values and Entrepreneurship: A Preliminary Consideration", *Journal of Enterprising Culture*, Vol. 3, No. 3, pp. 245–60.

Koontz, Harold (1961), "The Management Theory Jungle", *Journal of the Academy of Management*, Vol. 4, No. 3, pp. 174–88.

Leavitt, Theodore (1958), "The Dangers of Social Responsibility", *Harvard Business Review*, Vol. 36, No. 5, pp. 41–50.

Le Guin, U. (1974), *The Dispossessed*, Penguin, Middlesex, UK.

Lufkin, J. C. F. (1990), Foreword in *International Corporate Governance*, edited by J. C. F. Lufkin and D. Gallagher, Euromoney Books, London.

Marshall, Alfred (1920), *Principles of Economics*, eighth edition, Macmillan, London, first published 1890.

Marshall, Bruce (1991), *The Real World*, Houghton Mifflin, Boston.

Martin, H. (1974), "What Do Bosses Do?", *Review of Radical Economics*, Vol. 6, pp. 60–112.

Maruyama, Magoroh (1982), "Mindscapes, Workers, and Management: Japan and the U.S.A.", in *Japanese Management, Cultural and Environment Consideration*, edited by Sang M. Lee and Gary Schwendiman, Praeger, New York, pp. 53–71.

Mintzberg, Henry (1984), "Who Should Control the Corporation?", *California Management Review*, Vol. 27, No. 1.

Mises, L. Von (1966), *Human Action: A Treatise on Economic*, Henry Regnery Company, Chicago.

Morrison, Philip and Emily (eds) (1961), *Charles Babbage and his Calculating Engines*, Dover Publications, New York, a reprint from Charles Babbage (autobiography), Passages from the Life of a Philosopher, Long & Green, London, 1864.

Murayama, Motofusa (1982), "The Japanese Business Value System", in *Japanese Management, Cultural and Environmental Consideration*, edited by Sang M. Lee and Gary Schwendiman, Praeger Scientific, New York, pp. 89–116.

National Commission on the Environment (1993), *Choosing a Sustainable Future*, World Wildlife Fund, Washington, DC.

New Oxford English Dictionary (1986), Vol. 7, Oxford University Press, Oxford.

OECD (1975), *The Polluter Pays Principle: Definition, Analysis, Implementation*, Paris.

Pang, Y. H. (ed.) (1995), *Contemporary Issues in Accounting*, Addison-Wesley, Singapore.

Paton, W. A. and A. C. Littleton (1940), *An Introduction to Corporate Accounting Standards*, American Institute of Certified Public Accountants, Iowa City, IA.

Peck, M. J. (1988), "The Large Japanese Corporation", in *The US Business Corporation: An Institution in Transition*, edited by J. R. Meyer and J. M. Gustafson, American Academy of Arts and Sciences, Ballinger Publishing Company, Cambridge, MA, Ch. 2.

Peilly, William K. (1984), Foreword in *Are Environmental Regulations Driving U.S. Industry Overseas?* by H. Jeffrey Leonard, The Conservation Foundation, Washington, DC.

Peters, T. S. J. and R. H. Waterman (1982), *In Search of Excellence*, Harper & Row, New York.

Pierson, F. C. (1959), *The Education of American Businessmen: A Study of University-College Program in Business Administration*, McGraw-Hill, New York.

Pinchot, G. III (1985), *Intrapreneuring*, Harper & Row, New York.

Polak, Fred L. (1961), *The Image of the Future*, Vol. 2, Oceana Publications, New York.

Schollhammer, H. (1982), "International Corporate Entrepreneurship", in *Encyclopedia of Entrepreneurship*, edited by C. A. Kent, D. L. Sexton and K. H. Vasper, Prentice Hall, Englewood Cliffs, NJ, pp. 209–23.

Schumpeter, J. A. (1934), *The Theory of Economic Development: An Inquiry into Profits Capital, Credit, Interest and the Business Cycle*, translated by R. Poie, Harvard University Press, Cambridge, MA.

Seager, Joni (ed.) (1990), *The Corporate Way*, Simon & Schuster, New York.

Simon, Herbert (1967), *Economic Theory*, Vol. 3, Macmillan, New York.

Smith, Adam (1963), *The Wealth of Nations*, Vol. 2, Richard D.Irwin, Homewood, IL.

Smith, L. (1978), "The Boardroom is Becoming a Different Scene", *Fortune*, 8 May.

Steiner, George (1975), *Business and Society: Environment and Responsibility*, third edition, McGraw-Hill, New York.

Tann, Jennifer (1970), *The Development of the Factory System*, Commarket Press, London.

Techniques of Great Masters of Art (1985), Chartwell Books, New Jersey.

Timmons, Jeffry (1977), *New Venture Creation*, Irwin, Homewood, IL.

Turner, R. Kerry, David Pearce and Ian Bateman (1994), *Environmental Economics*, Harvester Wheatsheaf, Hertfordshire, UK.

Walter, Ingo (1973), *Environmental Management and the International Economic Order: An Agenda for Research*, D.C. Heath, Lexington, MA.

Williams, J. J. (1995), *Positive Accounting Theory: A Synthesis of Method and Critique*, Addison-Wesley, Singapore.

World Commission on Environment and Development (1987), Our Common Future, Report of the World Commission on Environment and Development, Oxford University Press, Oxford.

Index

period of economic transition, 150

Socialist governing system, 216

Tiananmen Square incident, 150

Chinese cultural values, 183–86
attitude towards environment, 185
business philosophy, 185
interpersonal relations, 184
national traits, 183
personal traits, 185
social (family) orientation, 184
work attitude, 184

civilization, 151, 175

classical management theory, *see* theories

Cold War, 21, 150, 369

collective bargaining, 126

collective contribution, 194

colonialism, 84, 113–14, 138

colonization, 114

command economy, 21

common good, 9–10, 14, 16, 20, 54, 121–23, 134, 136, 160, 179, 190, 193, 220, 244, 261, 344, 360

common ground, 164, 176, 180–81

common stock, of no-par value, 166
preferred, 171

common value, 181

communication, 33, 272, 282, 362, 363
lines of, 273–74
procedures, 277–79
regulations, 277–79

communication skills, *see* entrepreneurial skills

communism, 22, 121–23, 127, 133, 297

communist countries, 98, 369

communists, 131–32, 369

competition
elimination of, 84

competition, imperfect, 56

computerization, 174

Confucian ideas/theories, 62, 69, 287–90, 371

Confucian philosophy in business, 289–90

Confucian social conventions, 183

consequences of technology, 60

conservatism, 41

constitution, American, 14, 180

consumer behaviour, 143

consumer control, 11

consumer(s), 4, 116, 133, 141–42, 147, 293–94, 295, 364
democratically given right of, 84

consumers' recognition, 156

continental economy, 326–28

contractual agreement, 126, 192, 229

corporate behaviour, 100, 193

corporate by-law, 12, 88

corporate charter, 3

corporate citizen, 156

corporate commitment, 339

corporate control, 171

corporate culture, 109, 198
a simplified model for, 225
topline approach to, 226–27

corporate democracy, *see* democracy

corporate democratic rights, 172–73

corporate enterprising culture, 194, 198, 233, 293–94, 296
definition of, 195
three pillars of support, 231

corporate entrepreneurial environment, 284

corporate entrepreneurial model, 284

transition of ownership of, 164
Hornaday, J.A., 243, 245, 248
Howell, J.E., 77
human behaviour, 113
human economy, 174, 369
human empathy, 292
human interaction, 285, 351
human judgement, 304
human market behaviour, 62
human relations, 63
human resources policy, 170
human resources, *see* resources
human rights, 16, 150, 164,
   171, 180
humanities, 155
humanity, 19, 22, 118, 133,
   154, 354
  enemy of, 103
  future, 162
  life-support system of, 133,
   346
humanless environment, 351

I

imperialism, 84, 113–14, 138
income tax, 166
individual decision-making
  capacity, 134
individual goals, 60
individual identity, 181
individual knowledge, 161
individual skills, 161
individual(s), 126
  "foreign value" of, 174
  measurement of, 285
  motivated by natural instinct,
   125
  self-motivated, 94
individual's behaviour, 358
individual's own discipline, 296
individual's vision, 355
Industrial Accountants Society,
  80

industrial economies, capital
  driven, 111
Industrial Revolution, 61, 62,
  101, 137
industrial structure, 137
industrial waste, 110
industrialist, 127
inflexible hierarchy, 267
information highway, 174
information processing, 53
innovation, 53, 121, 290, 366
innovative desire, 124
innovative firms, 26
innovative process, 147
innovative schemes, 15
innovativeness, 25
inovalue, 251
International Council for Small
  Business (ICSB), 101
Internet, 94
interpersonal skills, *see* entrepreneurial skills
interpreneurship, 351
intrapreneurship, 279–80, 351
intrepreneurship, 54, 60, 351
Inuits and distribution systems,
  121–22
invention, success of, 7
inventor, 60
Inventory of Generally Accepted
  Accounting Principles for
  Business Enterprises, 37–38
investors, 109, 126, 127, 160,
  355
  see also stockholders
invisible hand(s), 2, 18, 20,
  343, 356
island economy, 326–28
"ism", 113–14, 118, 124

J

Japan, a commercially powerful
  nation, 105

economic success of, 181
Japanese business ethics "Shinto-ism", 182
Japanese business value system, 181, 358, 371
Japanese corporation, big family concept, 107
capitalistic expansion of, 105–7
devotion to the national entity, 182
employees as first priority, 182
lack of entrepreneurial challenge, 306
lifetime employment in, 107
robotics development, 107
technology transfer from, 106
Japanese culture, 181, 306
wisdom of, 306
Japanese management, 65, 306
Japanese mindscape, 66, 68
job creation, 123
Journal of Enterprising Culture, 199
"just" society, 121

K

Kandinsky, Wassily, 159
Keynes, John Maynard, 85
Kierstead, B.S., 117, 121, 132, 140
Kirby and Fan, 243, 245
knowledge barriers, 54
Kohl, Helmut, 151, 181, 370
Koontz, Harold, 85

L

labour market, *see* market
labour movement, 327
Laufer, Arthur C., 85
Lawrence, Rawl, 368–69
leadership, 73, 258–63, 280–82, 362
artificial, 263
earned, 261
solid, 263

statutory, 261
leadership types
autocratic or position, 259–60, 262, 263, 280–81
democratic, 259–60, 262, 263, 281–82
entrepreneurial, 259–62, 263, 282
professional or technical, 259, 262, 263, 281
leading without domineering, 259
League of Nations, 180
legalized stealing, 117
leverage, 167
liberal thinking, 95
life science, 148
Lincoln, Abraham, 182
"little dragons" team, 149
long-term environmental liability, 334–35
long-term relationship (between individual and corporation), 296, 298

M

macro-scanning, 143
macroeconomics, 20
macroeconomy, 308
Mahathir Mohamad, Dr., 106
management
bureaucratic, 63
by crisis (MBC), 71
by exception (MBE), 71
by fear (MBF), 58, 71, 280, 296, 351
by objective (MBO), 71, 296
defensive type, 211–12
education, 79
participatory, 69
personalized, 190
as practical functions, 91
as a profession, 76
professional, 77

scientific, 62, 76
strategic, 67
systemized, 62
Management Accountants
Society, 80
management committee, 172
management cultural differences,
8, 359
management culture
entrepreneurial, 199
traditional, 199
management decisions, 165
management focus, 296
management philosophy, 80, 82,
294–95, 296
management strategies, dehumanized, 54
management strategy, long-term,
343
management techniques, 76
management theories, *see*
theories
Management Theory Jungle,
The, 85
decision theory school, 86
empirical school, 85
human behaviour school, 85
management process school, 85
mathematical school, 86
social system school, 85
management trends, 59
market
financial, 128
free, 93
labour, 62
open, 119
market economy, 2–3, 18, 21,
43, 53, 76, 119, 127, 151,
155, 160, 176–79, 352
free, 20
open, 67
within a socialist governance,
98, *see also* China
market price, 165

market system, 176–77, 192
marketing strategy, customer-
oriented, 350
marketplace, 34, 122, 127, 148,
353
Marshall, Alfred, 44, 46–47
Marx, Karl, 65, 118, 128
Marxist-Leninist doctrine, 98
Maslow, Abraham H., 287
matrix form system, *see* organizational forms
measurement (of individuals), *see*
individuals
mechanism for economic
growth, 93
metaphysics, 162
micro–macro gap, 20, 308
microeconomics, 20
microeconomists, 20, 87, 88, 159
microeconomy, 308
mindset, 96, 163, 165,
238–241, 313, 361, 371
corporate managers', 22
entrepreneurial, 94, 236
mindset change, 240
minority shareholders, 172, 357
minority stockholders' rights,
171
Mintzberg, Henry, 11
Mises, 237
monetary unit, 176
money
as a common denominator, 43
as the medium of exchange, 43
"Motherhood", 180
Murayama, 370, 371

N

NAFTA (North American Free
Trade Agreement), 53, 152,
154
National Commission on the
Environment, 103
Nationalists, 131

NATO (North Atlantic Treaty Organization), 150
natural sciences, 145–49, 155
  biochemistry and chemistry, 147
  physics, 145–47
negotiation, simple guidelines to, 304
negotiation skills, *see* entrepreneurial skills
new venture creation attributes, 247
new venture creation strategy, 96
newly industrialized countries
  first generation, 14
  second generation, 14
Newton's fundamental laws of nature, 134–35
Next Step in Management Science, The, 85
non-economic social goals, 83
non-renewable natural capital (NRNC), 315–16
non-transferable, 169, 230
North American learning culture, 197
not-for-profit organization, *see* organization
nuclear power, 48
nuclear waste, 48

O

OECD (Organization for Economic Cooperation and Development), 313, 321
operating/operational residuals, *see* residuals
operating statement, 228
operational structure, 143
opportunity cost(s), *see* cost(s)
Organization for Economic Cooperation and Development, *see* OECD
organization(s)
  business, 265, 267

entrepreneurial corporate, 270, 281–86
entrepreneurial corporate, characteristics of, 283–85
not-for-profit, 29
organizational culture, 291, 293
organizational forms
  bureaucratic pyramids (factory system), 268, 271, 275–76, 279, 281, 363
  entrepreneurial corporate organization, 270
  entrepreneurial sub-contracting, 267–68
  flexible firm, 267, 269
  guild system, 267–68
  matrix form, 267, 269, 279
  profit centre, 270, 280, 363
organizational hierarchy, 271–72, 275
organizational illnesses, 363
organizational pyramid, 275–76
  and human relations, 277
organizational strategy, 265, 286
organizational theory, *see* theories
organized crime, 84
Owen, Robert, 63
owner control, 11
ownership, 55, 119, 123, 125, 127, 163–65, 215, 218–20, 222–25, 357
  bottomline for, 172
  China and Taiwan, 216–17
  corporate, *see* corporate ownership
  decision-making, 172, 233, 235
  employee participation in, 164, 170
  partial, 171
  private, 361
  quality of, 172
  right of, 116

human, 158
individual, 180–81, 192–93, 358
perceived differences between East and West, 186–87
personal, 161, 214, 290, 344
venture start-up, 124
visibility, 283–84
vision
  corporate, 140–46, 155–56, 355
  customer-driven, 355
  futuristic, 144
  global, 138
  profit-driven, 356
voting member, 229
voting power, 171

W

wage system, 116, 354
wars, 84, 360
Warsaw Pact, 21, 150
waste disposal, 365
water pollution abatement, 82
wealth 10, 163, 354
wealth creation, 7, 16, 21, 24, 25, 119, 133–34, 138, 156, 233, 351, 352, 363
wealth distribution, 118
Wealth of Nations, The, 20, 46, 66, 160, 197
Weber, Max, 63
Wee Cho Yaw, 297–98, 330, 364, 372

Whitehead, Alfred North, 158
wholly owned corporate subsidiaries, 88
wisdom, 81–82, 111, 193, 233, 253
  collective, 191
  conventional wholesale, 55
wisdom of democracy, 176
Wong Soke Yin, 239, 371
worker control, 11
workers, 164
  active, 128
  enterprising, 51
  propertyless, 116
  retired, 127
  unmotivated, 164
  working class, 127
World Commission on Environment and Development, 311, 365
World Trade Organization, *see* WTO
World War, Second, 132, 149, 215
WTO (World Trade Organization), 53, 151

Y

yuppie culture, 197

Z

Zhou En-lai, 182–83, 215